TANTRA PHILOSOPHY

TANTRA PHILOSOPHY

Editor-in-Chief
Swami Chidatman Jee Maharaj

ANMOL PUBLICATIONS PVT. LTD.
NEW DELHI - 110 002 (INDIA)

ANMOL PUBLICATIONS PVT. LTD.
H.O.: 4374/4B, Ansari Road, Daryaganj,
New Delhi-110 002 (India)
Ph.: 23278000, 23261597
B.O.: No. 1015, Ist Main Road, BSK IIIrd Stage
IIIrd Phase, IIIrd Block,
Bangalore - 560 085 (India)
Visit us at: www.anmolpublications.com

Tantra Philosophy

First Published, 2009
ISBN 978-81-261-3625-4

PRINTED IN INDIA

Printed at Mehra Offset Press, Delhi.

Contents

Preface

India colourful and vibrant, a land as diverse as its people. A mosaic of faiths, cultures, customs and languages that blend harmoniously to form a composite whole. One of the world's oldest living civilizations—which gave to the world—the concept of zero, the primordial sound Aum..., Yoga and Buddhism.

She is old. She is young. The two diametrically opposite, different yet unique and dynamic faces of India coexist even at this turn of the millennium. With a history of heritage, learning and civilisation dating back to almost 5000 years or perhaps even more, the modern India presents itself as an ideal country. From snow clad mountains to breezy sparkling blue sea side. From arid deserts with shifting sand dunes to lush green country sides. From ancient relics and architectural ruins dating back into antiquity to modern cities, India has it all.

To the north it is bordered by the world's highest mountain chain, where foothill valleys cover the northernmost of the country's maximum states. Further south, plateaus, tropical rain forests and sandy deserts are bordered by palm fringed beaches. Its southernmost tip is the meeting point of three Oceans.

India is a land of staggering contrast with a mingling of the tradition and modernity that is an unique experience to be in India. One will forever remember the stay in this wonderful Homeland-a land truly of mystery and enchantment. Only India can offer such an astonishing variety of contrasts.

She has given birth to numerous spiritual leaders and founders of some of the popular religions of the present day world. India probably has the most religious diversity in any country. It's the birthplace of Hinduism, Buddhism, Sikhism and Jainism and is among the few places to have a resident

Zoroastrian population. India with her veritable treasure trove of culture, mysticism, philosophy, art, music and architecture to name a few. Much of India's classical music is devotional and the North Indian Hindustani and South Indian Carnatic streams are distinct and both have a complex 'raga' framework. The legacy of dance in India is tremendous. The classical dances of India are numerous such as Kathakali of Kerala, Bharatnatyam of Tamil Nadu, Kuchipudi of Andhra Pradesh, Manipuri and Odissi from Orissa are the prominent dance forms in this country that sways to an altogether novel beat. The Yakshagana, nautanki and puppetry are ancient folk forms that live on till date. The earliest specimens of Indian painting are the ones on the walls of the Ajanta Caves dating back to 2nd century BC. The Mughals had a huge impact on Indian art. The influence of Persian art brought placid garden scenes, illustrations from myths, legends and history into Indian art. Word craft, handicrafts, architecture and sculpture all contribute to this rich and varied domain. Indian literature, both in English and in the vernacular, is ever more popular around the world.

The book aims at presenting perspectives of different aspects of Indian culture. It has topics on Indian classical music and dance, painting, ayurveda and yoga which are traditional and indigenous system of physical and mental wellbeing, language, philosophy and excellence achievements by Indian in the different branches of science and social sciences. It also includes our ancient religious and classic literature.

It is hoped that this book will be welcomed by the general readers as also scholars who are interested in the Indian heritage.

—*Editors*

1

Tantra in Indian History

The history of Tantra, as with that of most religions, is obscured by time. Many tantras offer mythical explanations for their origins, often setting themselves as the given word of either Shiva or a goddess such as Devi. Scholarly depictions of their origins are often as varied, ascribing tantras to pre-Aryan, Indus Valley civilizations or similar aboriginal, tribal groups or as integral part of an Indian cultural fabric. In reality, no definitive accounting of the origins of Tantra can be made owing to the significant polyvariance of the term Tantra in Sanskrit.

Tantra, which in its earliest written form was a distinctly iconoclastic, private, and esoteric practice, evolved into a number of respected, exoteric orders (sampradaya). It is convenient, although somewhat false, to group the orders into two categories: left-handed and right-handed. Left-handed tantras (vaama maarg) incorporate five sacraments (pancamakarapuja) of fish, meat, parched grain, wine and sexual intercourse into ritual practice. Right-handed tantras (dakshin maarg), on the other hand, advocate the visualization of these antinomian practices. Both groups rejected many aspects of Brahamanic orthopraxy, most notably the caste system and patriarchy. Despite this, Tantra was accepted by some high-caste Hindus, most notably the Rajput princes. Nowadays, Tantra has a large, though not always well-informed, following worldwide.

Hindu Tantra

Tantra exists in Vaishnava, Shaiva, and Shakta forms, among others. Extolled as a short-cut to Self-realization and spiritual enlightenment by some, left-hand tantric rites are often rejected as dangerous by most orthodox Hindus. The popular perception of Tantra among Hindus espoused in Indian journalism equates it with black magic.

Concepts and practices associated with Hindu Tantra:

1. The whole Universe is a creation of a Divine Consciousness (Shiva) and his active willpower (Shakti).
2. *Maya, the Illusory Nature of the Universe is Conceptualized as Maya Shakti:* the divine mother who causes humans to identify with illusory forms, preventing them from seeing the fundamental reality of their inseparability of the divine. Unlike in most orthodox Hindu traditions, Maya Shakti is held in considerable regard in Tantra and the attaining of ultimate realisation takes place in active concert with her.
3. *Embracing of Opposites:* This follows from the primary assumption that everything is a manifestation of the divine. Phenomena nominally considered impure and immoral are thus accepted and even utilised in transgressional practises, to further spiritual evolution.
4. *Non-acceptance of the Caste System, Class or Gender:* Tantra is widely recognized to be a system rejecting the orthodoxy of Brahmanism and the hierarchy of the caste system, accepting initiates from all backgrounds. However historical examples of elitism and perpetuation of casteist orthodoxy abound; each particular tradition must be viewed within its own historical and socio-political context in this regard.
5. *Importance of Guru:* The Guru occupies a godlike position in Tantra and the presence of a Guru is an essential condition for Tantric practise. Tantric practice is said to be a speedy but highly perilous route to enlightenment; unnavigable without the presence of an experienced master who will initiate the aspirant and guide him in his spiritual journey. An emphasis is placed on secrecy and the oral transmission of the tradition by the guru to a worthy pupil.

Tantra and the Vedas

Hindu Tantra is sometimes conceived as being antithetical to the more mainstream religious views of the Vedas. However there are commonalities suggesting considerable cross-influence. Also the ascribing of esoteric meaning and symbolism to the vedas is commonplace in the Tantric tradition. In Tantra the esoteric symbolism is said to be implict in the vedas, as well popular mythological tales like the Ramayana. Tantric concepts are said to be exquisitely mirrored in more exoteric myths, hidden to the uninitiated. The esoteric (tantric) interpretation often takes the form of meditational instruction where the body and its processes have a microcosmic correspondence with the macrocosm.

David Frawley cites one such instance. One Vedic verse which entreats "Offer Agni to Indra and Varun, so you may drink Soma", is interpreted as follows: Indra the King of Indriya is said to symbolize the senses. Varun which means Water/ Rain, also stands for Prana or life force. Agni refers to Kundalini. Soma, the feeling that makes one drunk with happiness occurs when kundalini is raised. The verse above is said to contain the following esoteric interpretation: "The person who has controlled his Senses (Indriya, which includes Sex) and his Prana (with Pranayama), only that person can raise the Kundalini and feel the Divine Bliss (Soma)".

In the Tripura Rahasya, one of the central texts of Shakta Tantra, says "This text has been created by summarising the teachings of the Vedas, Puranas and other Scriptures".

Tantra and Sex-traditional and Neo-tantric Views: To many Westerners Tantra is perceived as being an artful technique for enhancing sexual pleasure. It is also a common conception that it encourages free sex as a path to divine bliss. This reception in the west resulting from selective appropriation by Neotantrics, has met with vehement protests from Tantric practioners-themselves a heterogeneous group that include more conservative elements; some of them have sought to distance themselves from it, by attempting to symbolically represent sexual rituals or restrict their applicability.

According to Tantra one can go beyond the sexual plane of existence only by accepting it fully, utilizing it and spiritualizing ones innate sexual tendencies towards greater awareness. Several approaches are recognized as a way of conditioning aspirants prior to sexual meditation. Individual proclivities and spiritual achievement are taken into account by the guru from whom such training is sought. The path is usually considered appropriate only for certain individuals whose temperament and self control will enable them to forego sexual indulgence—a necessity, if the act is to serve as a tool to transcend identification with the mortal body. As with other tantric practices the presence of a guru is said to be a paramount and essential condition.

When done successfully as enjoined by the tantras it will culminate in an experience of infinite awareness. The tantric texts specify that sex has three distinct and separate purposes-procreation, pleasure and liberation. Frictional orgasm is eschewed by those seeking liberation for a higher form of ecstasy, as the couples are locked in a static embrace. Several sexual rituals are recommended and practised. These involve elaborate and meticulous preparatory and purificatory rites. The act results in a balancing of ida and pingala energies in both participants.

The shushumna nadi is awakened and kundalini rises up, resulting in an awesome cosmic awareness in which all residual individuality is completely disintegrated in the stream of samadhi consciousness. The act itself has several depths of meaning: the male and female participants are joined physically, representing Shiva and Shakti, the male and female principles. The act results in the subtle fusion of Shiva and Shakti energies resulting in an undivided energy field. In addition each participant experiences a fusion of their own Shiva and Shakti energies.

Neo-Tantra and New age advocates in the west have selectively co-opted tantric techniques by attempting to integrate them in an often eclectic way. These may serve the purpose of enhancing sexual pleasure, creating altered mind states or spiritualizing the sexual act. Some have attempted to demystify sexual rituals by stating physiological or psychological explanations. Neotantrics tend to divest their practice from ritualised conduct and religious imagery, emphasising a simple and secular approach. Western practitioners such as Julius Evola and Arthur Avalon agreed modern western interpretations were based on sensationalism and hedonism moreso than dharmic necessity implicated by the Kali Yuga.

Kundalini in Hindu Tantra: Kundalini is an important part of Hindu Tantra. According to Tantra, the Universe was created when The Supreme Consciousness, Shiva, felt a desire to multiply. A part of him was separated off and became his Shakti (Power). Shakti then began the creation of the universe, including the Heavens, Hell and the Earth. She reached a point where she could go no further. At the request of Shakti, Shiva enters her to infuse the worlds with consciousness. This embrace of Shiva and Shakti created the Universe. The whole Universe is a part of Shiva, and any part of it contains the entire universe. In our body, Shakti, the resides in the Muladhara, at the base of the spine. Due to Tamogun, she is asleep, and must be awakened. There are seven chakras (or energy centres in the body), which correspond to the Worlds Shakti created. As the persons Kundalini rises through the higher chakras, ever subtler and virtuous states are realised.

Right Hand and Left Hand Paths: Although a popular categorisation, the division into two paths is fairly artificial-though somewhat convenient—when Tantrik practices are described. A considerable admixture is present, depending on individual traditions and historical context. Many Tantrik traditions will fall between these twin polar approaches. Personal context, biases or attitudes of individual practioners may also dictate the predominant 'handedness' of individual paths.

The Right Hand Path consists of practises that use Bhakti coupled with Yogic exercises to raise Kundalini. They live lives that are said to be spiritually impeccable and the emphasis is on shunning worldly pleasures. Bhakti, or Divine Love is the key part in the Right Hand Path. The Tantric worships a Deity like Kali, Shiva, Tripura etc. with such intense love (and tools like Mantra, Yantra, visualisation etc.) that the Deity appears to the Sadhak, and starts guiding him/her.

The Left Hand Path is the one associated with the so called "darker" side of Tantra. Tantriks engaged in left hand practises embrace and accept what is usually considered repulsive to the senses, or what are nominally temptations to be avoided, such as sex and alcohol. A key tenet in Tantra is to accept everything as a manifestation of the divine. Thus mentally overcoming what the mind otherwise sees as repugnant, like bone, uncooked meat is an important practise on the path to master the mind. However the goals remain the same as those of any yogi; to overcome reactive elements of the mind and achieve complete control over it. This path, however, is seen as more treacherous and the presence of a guru, all the more important.

Relationship between Hindu, Buddhist and Chinese Tantra: Although Buddhist scriptures had arrived in Tibet as early as 173 CE, they were not properly 'tantricized' until 774 CE with the arrival of Padmasambhava who merged tantric Buddhism with the local Bon religion to form what we now recognize as Tibetan Buddhism. Chinese Tantra is similar to Tibetan, and was formed by the mixing of Chinese and Tibetan culture.

The three traditions interacted considerably. The Tibetan Master Saraha had a Shaivite Guru, while the Kashmiri Master Abhinavgupta had several Tibetan Gurus.

Another similarity between Hindu and Tibetan Tantra is their worship of Tara. However, while Tibetans worship generally her in her Kind form, the Hindus worship her in her wrathful form.

Japan's Shingon Sect: Buddhist tantrism is also practiced to a lesser extent in East Asia. Japan's Shingon sect, for example, practices tantric veneration of the deity Vairocana.

New Age Tantra or Neo-Tantra: New Age appropriations of Tantra usually disregard requirements involving guruparampara and ritual conduct, though they otherwise adopt many of the terms and concepts of Indian Tantra. In these circles, "Tantra" is often a synonym for sacred sexuality, a belief that sex ought to be recognized as a sacred act which is capable of elevating its participants to a higher spiritual plane.

Tantric Practices: Because of the wide range of groups covered by the term Tantra, it is hard to describe tantric practices definitively. The basic practice, the Hindu worship known as puja may include any of the elements below.

Mantra and Yantra: As in other Hindu and Buddhist yoga traditions, mantra plays an important part in Tantra for focusing the mind. The mantras used often invoke specific Hindu gods such as Shiva and Kali Ma. Similarly, puja often involves concentrating on a yantra or mandala associated with a deity as well.

Identification with Deities: Tantra, being a development of early Hindu-Vedic thought, embraced the Hindu gods and goddesses, especially Shiva and Shakti, along the Advaita philosophy that each represents an aspect of the ultimate Para Shiva, or Brahman. These deities may be worshipped externally with flowers, incense, and other offerings; but, more importantly, are used as objects of meditation, the practitioners either visualizing themselves as the deity or experiencing the darshan (vision) of the deity visualized as appearing in front of the pratitioner. In Buddhist Tantra, this process is known as deity yoga.

Concentration on the Body: Tantrikas generally see the body as a microcosm; thus in the Kaulajnana-nirnaya, for example, the practitioner meditates on the head as the moon, the heart as the sun and the genitals as fire. As in the yoga tradition, a series of energy centres (chakras-"wheels") may be used as concentration points, and may be associated with elements, planets or occult powers (siddhi).

Taboo-breaking: The act of breaking taboos is the definitive feature of left-hand Tantra. Many tantras require specific preparations or purifications prior to breaking taboos. The normal state of human preparation is referred to as pasu-bhava (animal disposition). A person in the state of pasu-bhava is one who regularly eats meat and indulges in intoxication. They are considered unprepared for tantric practice. Tantras prescribe a strict regimen of purification before an individual moves onto the actual practice. An individual who successfully completes this purification may take a vow of viravrata (a hero's vow) to be of vira-bhava (heroic disposition). They may then engage in the practices which involve the breaking of taboos.

In the Kaula and Vamachara schools of Tantra the panca makara, or 5 Ms, is a ritual sacrament in which several Hindu taboos ritually broken in order to free the practitioner from binding convention. The five substances involved are:

- madya (wine)
- mamsa (meat)
- matsya (fish)
- mudra (parched grain)
- maithuna (intercourse)

The "sacramental" or ritual breaking was only for the vira practitioner, not the divya or pasu. The pasu would misunderstand and get caught up in the literal act while the divya will have already progressed beyond and not need the literal act to understand the inner meaning.

The schools which practice right-hand Tantra substitute innocuous items for the taboo substances and acts, claiming that literal interpretations of the panca makara miss the inner truth of the rite.

Neo-Tantra

Neo-Tantra refers to modern or otherwise New Age appropriations of Tantra that usually disregard requirements involving the guidance of a guru and ritual conduct, though they otherwise adopt many of the terms and concepts of Indian Tantra. In these circles, "Tantra" is often a synonym for sacred sexuality, *i.e.* a belief that sex ought to be recognized as a sacred act which is capable of elevating its participants to a higher spiritual plane. At other times "Tantra" comes to mean a set of techniques for cultivating a more fulfilling sexual or love relationship, like the otherwise unrelated Kaama Sutra, regardless of its mystical import. Bhagwan Shree Rajneesh, later known as Osho, used Tantra in combination with breathing techniques, bioenergy, yoga and massage in some of his groups. His students continued this concept and developed it. One of his successful and well known students is Margot Anand. Tantramassage (Tantric Massage) is a form of massage developed in the 1980s by Andro, Andreas Rothe, a teacher for massage, meditation and yoga in Berlin, Germany. This method became very popular in Germany and Switzerland.

Many other teachers have pioneered their modern understanding of Tantra as it relates to its ancient roots. Another way of looking at is a top-down approach. In other words, tantric knowledge or teachings can be directly transmitted from teacher to student through the modalities of eye contact, playful conversation, touch and sexual intercourse.

Criticisms: Georg Feuerstein, a Buddhist who also trained under Hindu Tantra, writes in his book "Tantra: Path of Ecstasy",

> *"Many are attracted to Neo-Tantrism because it promises sexual excitement or fulfilment while clothing purely genital impulses or neurotic emotional needs in an aura of spirituality...Today translations of several major Tantras are readily available in book form, and many formerly secret practices are now, in the language of the texts, 'like common harlots'. This gives would-be Tantrics the opportunity to concoct their own idiosyncratic ceremonies and philosophies, which they can then promote as Tantra."*

Tantramassage: Tantramassage (Tantric Massage) is a form of massage developed in the 1980s by Andro, Andreas Rothe, a tantric "bodywork" teacher in Berlin, Germany.

His method includes various massage techniques, elements from Yoga and Sex therapy. The word Tantra in the name is meant to allude to a connection with an Indian cultural background Rothe claims to draw from.

Important influences came from the "Taoist Erotic Massage" by Joseph Kramer, Ph.D. and Annie Sprinkle. Joseph Kramer, founder of The Body Electric School and the new online erotic massage school The New School of Erotic Touch, emphasizes conscious breathing and skilled erotic touch in combination to wake up the body, generate erotic energy and create prolonged states of full-body arousal.

Other teachers in the field include Wilhelm Reich, Alexander Lowen and Mantak Chia.

Clients are in a passive role, supported by breathing techniques. Sessions last 90 minutes or longer. The Tantramassage movement is popular with New Age followers in Germany and Switzerland, some practitioners are organized in a tantric massage association (Tantramassage Verband e.V.).

Tantramassage claims not to engage in prostitution. Its adherents claim tantramassage does not include any kind of sexual exchange.

Ananda Marga: Ananda Marga, officially known as Ananda Marga Pracharaka Samgha (AMPS) meaning "the organization for the propagation of the path of bliss" was founded in Jamalpur, State of Bihar, India in 1955 by Prabhat Ranjan Sarkar (1921-1990), known by his spiritual name of Shri Shri Anandamurti. Ananda Marga bases its ideology on the theory that total human development can only be achieved through both personal development, and social service to the community.

The meaning of Ananda Marga is "path of bliss", ananda meaning bliss, and marga path. Through a system of meditation techniques, yoga postures (asanas), spiritual gatherings, and social service the ananda margiis (followers of AM) strive to develop themselves as human beings, and the betterment of others.

History

Early History: The early history of P.R. Sarkar and his way towards the establishment of Ananda Marga Pracharaka Samgha is obscure and few writings can be found. There is, however, one account which proved to be significant in relation to Sarkars future as a spiritual leader, namely the initiation of a notorious criminal named Kalicharan. In 1939, Sarkar left Jamalpur for Calcutta to attend Vidyasagar College of the

University of Calcutta. He frequently visited the ghats along the river Ganges at night time to meditate and contemplate. At that time of day ordinary people shun the ghats because they were frequented by thieves and other petty criminals.

One night (probably in 1939 or 1940) young Sarkar was performing his meditation, when suddenly assaulted by a notorious criminal and killer, Kalicharan. Kalicharan demanded him to hand over his wallet or be killed. At that point Sarkar, who didn't know the unknown assailant, is to have said: "Kalicharan, I promise to give you all my money, but first tell me whether you rob people out of necessity or habit?". Kalicharan realized that he was dealing with an extraordinary person, and decided to receive initiation into meditation from Sarkar. He asked what he could call his new spiritual master, who simply replied: "call me Baba", which became the name all his followers in the future would affectionately call him. From that time on Kalicharan changed his ways and got the spiritual name Kalikananda.

Recent History: Ananda Marga Pracharaka Samgha was officially established in 1955 in Jamalpur, Bihar, India. P.R. Sarkar was at that time working as a railway official in the Indian Railways. At that time he was a rather personal teacher having only a handful of spiritual students. However, the establishment of Ananda Marga was not always Sarkar's intention. In his book Ananda Katha, Acharya Nagina describes Sarkar's frustration with a world of ignorance and violence, and eventually his decision to not continue his spiritual work. After pleas from his most sincere followers, Sarkar eventually decided to form the AMPS and took a plea to lead humanity to a higher state of existence.

In 1959, the Progressive Utilization Theory (PROUT) was released. Claims have been made that this put Ananda Marga at stakes with the local communist authorities.

In 1962, a monastic order was formed, where the junior monks and nuns where known as brahmacharya, and the seniors known as avadhutas, where the former are dressed in a saffron coloured top and white bottom, while the seniors are dressed in full saffron robes.

In 1963, the Education, Relief, and Welfare Section (ERAWS) of Ananda Marga was formed. Schools, orphanages, and emergency relief work was being conducted under its name.

In the mid-1970's, Ananda Marga was banned by the Indian government of Indira Gandhi together with most opposition parties and Shri Prabhat Ranjan Sarkar was falsely charged with murder. He was initially found guilty and sentenced to lifelong imprisonment, but after a retrial after the Congress party lost power found him innocent. During his stay in prison, Shri P.R. Sarkar was administered poison by the jail authorities on the pretense that it was medicine. Surviving the attempt, Shri P.R. Sarkar then fasted for the next five years until his release out of protest of being denied a proper judicial inquiry into the incident. During his stay in prison his mission spread to all continents of the world.

Before 1971 AMPS was basically an Indian organization with few workers abroad, but during Sarkar's imprisonment his workers travelled the world and established local Ananda Marga centres in more than a hundred countries. What was once a national organization was now global.

P.R. Sarkar travelled to many countries after his exoneration, and worked continuously to spread his words. In 1982, he took up composing, and till his death in 1990 he managed to compose 5018 songs, which were called Prabhat Samghiit.

In 1986, he formed the theory of microvita, small, subtle entities which were said to be the foundation of all existence. Research on the field is being conducted by the Microvita Research Institute,, but their progress and findings are still limited.

On 21st of October 1990, he entered his maha samadhi and departed his physical body.

In the late 1990's and early 2000's, the state of the organization has been in disarray. Internal quarrelling and disagreements has led to a split. There are two fighting fractions, both Indians. One is mainly composed of Hindi speakers from Bihar, and the second Bengali speakers from Bengal. The origin of the conflict is uncertain, but one claim is that the Hindi fraction expresses inferiority compared to the Bengali fraction. The Bengali fraction is thought to have been closer to Sarkar during his life, and have held higher positions in the organization. The conflict is ongoing, and does not seem to have an early solution.

Organization: Ananda Marga is a hierarchal organization lead by senior members of AMPS, most of them dedicated monks and nuns. A well developed hierarchy is working in the AMPS organization where the Global Office is the main representative organ of AMPS, lead by the General Secretary. Globally the organization is divided into 9 sectors named after a city in each of them, namely:

1. Delhi Sector—the Indian subcontinent
2. Hong Kong Sector—North East Asia
3. Manila Sector—South East Asia
4. Suva Sector—Australia, Pacific region
5. New York Sector—North, and Central America, the Caribbean
6. George Town Sector—South America
7. Berlin Sector—Europe
8. Qahira Sector—Balkans, West Asia, North Africa
9. Nairobi Sector—Sub-Saharan Africa

Each sector is subdivided into regions, which again is further divided as per needs. Each of them is lead by a senior official, namely the Sectorial Secretary, Regional Secretary, and so on.

A separate entity of AMPS is the Women's Welfare Department (WWD), destined to help the betterment of women through education, social work, and local initiatives. WWD was started as a reaction to the common mistreatment and lack of respect women are suffering in many parts of the world. WWD is lead by women, for women, and one of their goals is the equality of both sexes.

AMPS also strives for development of the human sphere in other parts of life as well. Renaissance Artists and Writers Organization (RAWA) is an organization for the development of the fine arts. RAWA regularly arrange cultural events in several countries, those being concerts, theatrical acts, etc.

The official social welfare and development organization below AMPS is Ananda Marga Universal Relief Team (AMURT). AMURT arranges both emergency relief during natural disasters, and long term projects such as building of schools and orphanages.

Practice

Spiritual Teachings and Practice: Ananda Marga (AM) is the practice of AMPS. Ananda Marga meaning the path of bliss. AM is based on Tantra yoga, as interpreted by Shri Shri Anandamurti in the books "Discourses on Tantra volume 1 and 2".,. "Tantra" meaning the liberation from darkness, the root "tan" meaning darkness, and "tra" liberation.

Tantra in the popular Western sense meaning sexual practices is not a part of the Ananda Marga Tantra tradition. Sarkar in his book on Shiva, "Namah Shivaya Shantaya", and "Discourses on Tantra volume 1", he explains the difference between the lefthand crude tantric path, and the right-hand subtle path by giving an explanation of panchamakara, also known as the 5 Ms. The practitioner of Ananda Marga Tantra yoga sees the mind as a "trapped monkey" striving to escape from its cage.

The Tantra path is the path of the brave, because it requires the practitioner to follow strict rules as to their spiritual practice. Meditation is the main practice of this tantric tradition, and through the meditation the practitioner struggles to overcome his weaknesses and imperfections. Through daily meditations the practitioner faces his minds deepest secrets, and to proceed in the practice he must manage to overcome whatever he faces in the deepness. Because the path is considered so difficult many people stray from the path and reject the teachings afterwards.

The base Ananda Marga practice is covered in a set of rules called the sixteen points. The sixteen points guides the practitioner of the tantric path on both spiritual and social aspects.

Meditation: In the tantric tradition of Ananda Marga the spiritual aspirant is called a sadhaka, and he practices sadhana. Sadhana comes from the Sanskrit root word sadh which means "to complete". Sadhana signifies the effort through which a person becomes completely realized.

In the tantric tradition the spiritual master, the guru plays a special role. The guru, or dispeler of darkness, guides and leads his students on the razor sharp path on which he follows when he decides to become a spiritual aspirant. It is also stated in the tantric tradition that the student doesn't find the teacher, but the teacher finds the student.

When the student decides to aspire on the path of bliss he will be initiated by a qualified meditation teacher called acharya, Sanskrit for teacher. An acharya is most commonly a monk or nun, but there are also a few family acharyas in the Ananda Marga tradition.

In the initiation the aspirant makes a commitment to practice meditation, and is then taught the technique itself. The aspirant is then required to keep all his practices secret and not discuss them with others.

The base Ananda Marga meditation system is called sahaja yoga, meaning easy yoga, in the sense that everyone can do the practice. The sahaja yoga system consists of 6 meditation techniques called lessons. The lessons are taught one by one, on a personal basis. There is no specific system as to when one can learn a new technique, but is rather taught on a personal basis depending on the level of interest and dedication of the student. Thus some students learn all six lessons in a year or two, while some complete all lessons in over 20 years. There is also a higher set of meditation lessons taught by Ananda Marga, but these are highly secretive, and only a select few advanced practitioners are ever taught them.

Yoga and Physical Exercises: The physical practice of Ananda Marga comprises both yoga asanas, and two specific dances, kaoshiki and tandava.

Yoga Asanas: The Ananda Marga system of yoga comprises of 42 yoga asana postures, which were specifically selected by

P.R. Sarkar for their benefits. In the Ananda Marga system yoga asanas should be performed twice daily as per the following system, once in the morning, and once in the evening. After the practice a full body massage and final relaxation is performed to complete the asana practice.

Kaoshiki: Kaoshiki is a dance performed by all students, and consists of 18 mudras, each with a specific meaning. The dance is both a spiritual and physical exercise. The spiritual ideation of the dance is done to establish a link to the divine, and the physical strengthens and softens the body. There is also a claim that the exercise strengthens the nervous and endocrine system to give full body balance.

Tandava: This is a vigorous dance associated with Shiva as the cosmic dancer Nataraj. The name tandava is derived from the Sanskrit word tandu, which means to jump. This dance is only performed by the male followers of Ananda Marga because of its testosterone producing effect. The dance is performed to strengthen the practitioner and dispel all fear, even fear of death itself, and therefore the dance also has an ideating factor. The dancer starts off with the two arms outstretched, the left arm with an open palm, and the right arm with a clenched fist. The dancer ideates holding a human skull in the left hand, symbolizing death, and ideates holding a knife in the right, symbolizing the fight against all fear. The dance starts with a vigorous jump and landing in a position of bent knees. Another jump follows, and the dance continues in a jumping manner lifting one leg then the other continuously. The dance is ended with a final jump.

Activities: Ananda Marga runs yoga and meditation centres, schools, orphanages, food distribution centres, disaster relief programmes, medical centres, community development projects, and other service projects. Emphasis is placed on solving the problems of the local people, whatever they may be, and assisting them in managing all their personal and social resources for the prosperity of all.

Proclaimed Philosophy and Goals: The philosophy of Ananda Marga is one of universalism. It is a synthetic outlook, recognizing God as the one limitless supreme consciousness.

This universe, having been created by God, is existing in God, and as such all created beings are God's children, part of the one universal Cosmic family.

Thus we are all brothers and sisters, connected in love by the bonds of Cosmic fraternity. This universe of diversity has evolved out of unity (God), and we must continually strive to see that unity in all diversity. Thus the social outlook of Ananda Marga recognizes that the welfare of the individual is inextricably linked with the welfare of the collective, each relying on the other for its existence and dynamism. Everyone has the right to equal opportunity and as such there should be no discrimination on the basis of superficial barriers such as race, nationality and religion.

Ananda Marga advocates a world of justice, security and peace for all. To this end Ananda Marga gives a pratised, rational, and systematic way of life for the balanced development of all human potentialities: physical, psychic and spiritual. It is a system that incorporates practices that are beneficial for personal and social upliftment. These range from proper hygiene and diet, to yoga postures, to a scientific technique of meditation based on morality and leading to complete peace and inner fulfilment. It recognizes that a balance is needed between the spiritual and mundane aspects of existence, and that neither one should be neglected at the expense of the other. Hence, the goal of Ananda Marga is "Self-realization and the welfare of humanity."

Symbology

Pratik—The Symbol of Ananda Marga: Pratik pratisedly means emblem, and summarizes the ideology of Ananda Marga in a visible form. The pratik is composed of two interconnected triangles where one points upwards and the other points downwards. In the centre of the triangle a rising sun is placed, and in the core of the rising sun a svastika (counter clockwise) is seen.

Interpretation:

1. *Upwards Pointing Triangle:* Signifies work in the world, *i.e.* social work.

2. *Downwards Pointing Triangle:* Signifies inner work, *i.e.* meditation.
3. *The Sun:* The sum of work in the world and inner development leads to personal and spiritual development.
4. *The Svastika:* Signifies personal victory, in the sense of spiritual fullfilment and salvation.

Flag: A triangular, orange/saffron coloured flag with a white counter clockwise svastika in the centre is used by the Ananda margiis.

Relation to Hinduism: It is debatable whether Ananda Marga is a Hindu organisation. Only two of the traditional Hindu gods are recognised as Ishvara (the rest are considered to be mythological creations or elaborations), the caste system and the concept of religion as well as avatarism are rejected.

Criticism: Ananda Marga has faced some criticisms. As a political protest some members reportedly practiced self-immolation during the period when they faced repression in India. A report by the Italian police placed them alongside Soka Gakkai International for political activities. There are news agencies that published critical articles like the BBC-: article mentioning an arms drop in India-and The Australian-: article on possible involvement of Ananda Marga members on the Sydney Hilton bombing. Some have labelled Ananda Marga a terrorist organisation.

Supporters have pointed to events like (Ananda Marga members charged of Hilton bombing were released and pardoned) and (to whom the weapons were being delivered is still disputed), arguing that the criticism is based on speculations or unproven or disputed facts.

Various Aspects of Tantra

The Relationship between Buddhist Tantra and Hindu Tantra: Although some scholars have maintained that Buddhist Tantra was derived from Hinduism, this is not correct. The theory, prevalent among those who adhere to the tenets of the Hinayana, is based on a superficial resemblance of various elements of the two systems, such as the forms of the deities, the meditations on psychic veins and airs, the fire rituals, etc. Though certain practices, like the repetition of mantras, are common to both Hindu and Buddhist tantric traditions their interpretation, *i.e.* the inner meaning, is vastly different. Furthermore, Buddhist Tantra is superior because, unlike Hinduism, it contains the three principal aspects of the Path: renunciation, the enlightened attitude and the right philosophy.

To elaborate: as even animals want freedom from suffering, there are non-Buddhist practitioners who wish to be free from contaminated feelings of happiness and so cultivate the preparatory state of the fourth absorption (Dhyana). There are even some non-Buddhists who temporarily renounce contaminated feelings of happiness and attain levels higher than the four absorptions.

However, only the Buddhists renounce all these as well as neutral feelings and all-pervasive suffering. Then by meditating on the sufferings together with their causes, which are mental defilements, they can be abandoned forever. This is why, while non-Buddhists meditate on the form and formless states and attain the peak of worldly existence, samadhi, they cannot abandon the mental defilements of this state. So, when they meet with the right circumstances anger and the other passions develop, karma is created and the wheel of the circle of rebirth begins to turn.

Because of this and similar reasons, such practices are not fit to be included in the Mahayana. They resemble neither the common sutra path comprising: the attitude of renunciation which wishes for freedom from the cycle of rebirths; the wisdom which correctly understands egolessness, which is the right philosophy acting as an opponent to ignorance—the root of cyclic existence; and the development of the mind which aims for complete enlightenment for the sake of all sentient beings; nor do they resemble the practices of the exclusive tantric path of the Great Vehicle.

The Origin of Tantra: The tantras were spoken by the Buddha himself in the form of his supreme manifestation as a monk, also as the great Vajradhara and in various manifestations of the central deity of specific mandalas. The great beings, Manjushri, Samantabhadra, Vajrapani and others, urged by the Buddha, also taught some tantras.

In terms of the four classes of Tantra, the Kriya tantras were taught by the Buddha in the form of a monk, in the realm of the thirty-three gods on the summit of Mt. Meru, and in the human world where Manjushri and others were the chief hearers.

The Pung-Zang tantras were taught in the realm of Vajrapani. Others were taught by the teacher, Buddha himself, and with his blessings some were explained by Avalokiteshvara, Manjushri and Vajrapani while others were spoken by worldly gods.

The Carya tantras were also taught by the teacher Buddha in the form of his supreme manifestation in the celestial realms and in the realm called Base and Essence Adorned with Flowers.

The Yoga tantras were taught by the Enlightened One when he arose in the form of the central deity of each mandala in such places as the summit of Mt. Meru and in the fifth celestial realm of desire.

The Anuttara tantras were also taught by the Buddha. In the land of Ogyan the Buddha, having manifested the mandala of Guhyasamaja, taught King Indrabodhi this Tantra. The Yamantaka tantras were taught by the teacher Buddha at the time of the subduing of the demonic forces and they were requested by either the consort of Yamantaka or by the consort of Kalacakra. The Hevajra Tantra was taught by Lord Buddha when he arose in the form of Hevajra in the land of Madgadha at the time of destroying the four maras.

The Tantra was requested by Vajragarbha and by the consort of Hevajra. Having been requested by Vajra Yogini, the Buddha, in the manifestation as Heruka on the summit of Mt. Meru, taught the root Tantra of Heruka and, when requested by Vajrapani, taught the explanatory Tantra. As for the Kalacakra Tantra, the mighty Buddha went south to the glorious shrine of Dharnacotaka and there, manifesting the mandala of the Dharmadhatu speech surmounted by the mandala of

Kalacakra, taught this Tantra to King Chandrabhadra and others. Although he appeared in many different manifestations, actually the tantras were taught by the enlightened teacher, Lord Buddha.

What Happens during an Initiation: In the initiations of each of the four classes of Tantra, there are many differences, some great and some small, and so therefore one initiation is not sufficient for all mandalas. At the time of initiation some fortunate and qualified disciples, when receiving the initiation from a qualified master, develop the wisdom of the initiation in their mind streams. Unless this happens, sitting in initiation rows and experiencing the initiations of the vase and water, etc. will implant instincts to listen to the Dharma but little else. An initiation is necessary to study Tantra because if the secrets of Tantra are explained to someone who has not received initiation, the guru commits the seventh tantric root downfall and the explanation will be of no benefit whatsoever to the mind of the disciple.

The Relationship between Sutra and Tantra

Regarding renunciation and bodhicitta, there is no difference between Sutrayana and Tantrayana, but regarding conduct there is. Three kinds of conduct have been taught: the disciple who admires and has faith in the Hinayana should separate himself from all desires; the disciple who admires the Mahayana should traverse the stages and practice the perfections; while he who admires the deep teachings of Tantra should work with the conduct of the path of desire. From the point of view of the philosophy, there is no difference in emptiness as an object of cognition but there is a difference in the method of its realization.

In the sutra tradition the conscious mind engages in meditative equipoise on emptiness, while in Tantra the innate wisdom, an extremely subtle mind, is involved and the difference therefore is great. The main practice of Sutrayana, engaging in the path as a cause to achieve the form body and wisdom body of a buddha, is the accumulation of wisdom and virtue for three countless eons and the accomplishment of one's own buddhafields. Therefore, Sutrayana is known as the causal vehicle. In Tantra one concentrates and meditates, even while still a beginner, on the four complete purities which are similar to the result-that is, the completely pure body, pure realm, pure possessions and pure deeds of an enlightened being. Thus Tantra is known as the resultant vehicle.

The Four Traditions: As for the sutra tradition, the explanation of the Hinayana and Mahayana is the same in all the four great traditions. Also, as far as the preliminary practices are concerned, there are no differences apart from the names. In the Gelug tradition they are called the Stages of the Path of the Three Motives; in the Kargyü they are known as the Four Ways to Change the Mind; the Sakya refer to Separation from the Four Attachments; while the Dri-gung Kargyu speak of the Four Dharmas of Dag-pa and the Five of Dri-gung.

In Tantra, the individual master's way of leading the disciples on the path depends on his experience and the instructions of the tantric root texts, together with the commentaries of the great practitioners. These result in the entrance into practice being taught a little differently. However, all are the same in leading to the final attainment of the state of Vajradhara.

Tantric Symbols

For the untrained, especially the symbolism in Tantra can be extremely confusing. However, it should be noted that in modern psychology, Freud and Jung have clarified many aspects of the sub-consciousness in terms of symbolism. In Buddhism, something like sub-consciousness is an impossibility by definition-an awareness without consciousness does not make sense, but there are certainly areas of our mind we are only barely aware of. In order to access these more hidden and subtle aspects of our mind, symbols can be very effective in mind transformation.

As Jean Shinoda Bolen writes in The Tao of psychology:

> *"Jung describes archetypes as 'patterns of instinctual behaviour'.... One definition of archetypes that Jung uses refers to 'primordial images', or archetypal figures*

that become activated and then clothed with personally derived emotional coloration. This occurs when an emotional [or spiritual] situation develops that corresponds to a particular archetype... For example the 'Wise Old Man', the 'Divine Child', 'All-Giving Mother', 'Patriarchal Father', 'Temptress' or 'Trickster'-all are symbolic, recurring figures in dreams, literature and religions. When the archetypal level of the collective unconscious is touched in a situation, there is emotional intensity as well as a tendency for symbolic expression. Then the usual everyday level of experience becomes altered; there is more 'magic' in the air. One can become 'inspired' or be on a 'crusade'."

The very last word of this quote is important in the entire concept of Tantra; it can be efficient in enhancing spiritual progress, but if used unskilful, it can lead to madness like personal 'crusades' and 'holy wars'. This presents another good reason for the traditional secrecy of tantric practice and reliance on a true spiritual master.

Another important aspect is the fact that the Buddha clearly explained that "meaningless ritual" should not be practised. So by definition, one could say that ritual in Buddhism must be filled with (symbolic) meaning. Most of below symbols are taken from the Tibetan traditions, as they can be considered to have preserved the most complete set of tantric teachings.

The Vajra: The Vajra is a very important symbol in Buddhist Tantra. In fact, the Tantra teachings are even referred to as the Vajrayana or Vajra-vehicle. The vajra is probably a derivation of a weapon and a sceptre, and may have its origins in the trident and a mendicants' staff, and symbolises being indestructible, therefore sometimes compared to a diamond.

To give an impression of the vast symbolic meaning of many objects used in Tibetan Buddhist Tantra, below is a summary excerpt from the excellent book "The Encyclopaedia of Tibetan Symbols and Motifs" by Robert Beer.

"At the centre of the vajra is a flattened sphere representing the dharmata as the 'sphere of actual reality'. This sphere is sealed within by the syllable HUM, whose three component sounds represent freedom from karma (Hetu), freedom from conceptual thought (Uha) and the groundlessness of all dharmas (M). On either side of the central hub are three rings [which] symbolise the spontaneous bliss of Buddha nature as emptiness, signlessness and effortlessness. Emerging from the three rings on either side are two eight-petalled lotuses.

The sixteen petals represent the sixteen modes of emptiness. The upper lotus petals also represent the eight bodhisattvas, and the eight lower petals, the eight female consorts. Above the lotus bases are another series of three pearl-like rings, which collectively represents the six perfections of patience, generosity, discipline, effort, meditation and wisdom. A full moon disc crowns each of the lotuses, symbolising the full realisation of absolute and relative bodhicitta Emerging from the moon discs are five tapering prongs, forming a spherical cluster or cross.

The four [outer] curved prongs curve inwards to the central prong, symbolising that the four aggregates of form, feeling, perception and motivation depend upon the fifth aggregate of consciousness. The five upper prongs of the vajra represent the Five Buddhas (Akshobhya, Vairochana, Ratnasambhava, Amitabha and Amogasiddhi), and the unity of their five wisdoms, attributes and qualities. The five lower prongs represent the female consorts of the Five Buddhas (Mamaki, Lochana, Vajradhatvishvari, Pandara and Tara) and the unity of their qualities and attributes. The Five Buddhas and their consorts symbolise the elimination of the five aggregates of personality. The ten prongs together symbolise the ten perfections (the six mentioned above plus skilful means, aspiration, inner strength, and pure awareness); the 'ten grounds' or progressive levels of realisation of a bodhisattva; and the ten directions.

Each of the outer prongs arise from the heads of Makaras (sea monster). The four Makaras symbolise the four immeasurables (compassion, love, sympathetic joy and equanimity); the four doors of liberation (emptiness, signlessness, wishlessness and lack of composition); the conquest of the four Maras (emotional defilements, passion, death, divine pride and lust); the four activities or karmas; the four purified elements (air, fire, water, earth); and the four joys (joy, supreme joy, the joy of cessation and innate joy).

The tips at the end of the central prong may be shaped like a tapering pyramid or four-faceted jewel, which represents Mount Meru as the axial centre of both the outer macrocosm and inner microcosm. The twin faces of the symmetrical vajra represent the unity of relative and absolute truth."

The above describes the most common version of the five-pronged vajra. There are also the one-pronged, three-pronged and nine-pronged vajra, the double vajra, etc., each with their own extensive symbolism....

The Bell: In tantric rituals, the Vajra is normally held in the right hand, and the Bell in the left. In this combination, the Vajra symbolises method, bliss and male aspects, and the Bell symbolises wisdom, emptiness, the female aspect. The bell as a whole symbolises the sound of the Dharma and is often used in tantric rituals to offer sound.

The Bell is traditionally topped by a half-vajra (the 5 spokes are the five forms of mystical wisdom), and below is the face of Viarocana-the incarnation of universal truth (Dharma). On the shoulder of the bell are 8 lotus petals with syllables in them; the lotus leaves represent the 8 great Bodhisattvas and the syllables are their consorts. As you may suspect by now, each and every detail of the bell has an elaborate symbolic meaning. The whole of the bell can even symbolise a complete mandala.

Whenever one holds these objects, one should try to remember these symbolic meanings and gradually familiarise the mind with the idea; this is the main idea behind meditation.

Some other Tantric Implements: An implement carried around by most Tibetans is the Mala or a rosary of prayer beads. These are not unlike the Christian rosary, or the beads used in Islam and Hinduism. They are used to focus ones' mind on the recitation of mantras, and to count them as part of a practice.

The story of the beads' origin is as follows: "Sakyamuni, the founder of Buddhism, paid a visit to king Vaidunya...Sakya directed him to thread 108 seeds of the Bodhi tree on a string, and while passing them between his fingers to repeat... 'Hail to the Buddha, the law, and the congregation'... (2,000) times a day (Dubin)" (from this page not all the info here is correct though).

There are for example practices for which one is required to recite 100,000 mantras; obviously a simple counter is needed to keep track of this huge number. The Tibetan mala usually has at least 108 beads-this number probably originates to the 108 names for Hindu deities (incidentally, the same number is used in Islam to refer to God).

Tibetans often attach strings to their malas which have little sliding rings on them, these are to keep count of the number of malas; in such a way one can count up to 10,000 or ever more on one mala.

A very special object within Tibetan tantric practice is the Kapala or skullcup. The skullcup is related to detachment and transformation of the world we experience.

Often used in Tibet are the small handdrum (dhamaru) and the larger Chod-drum. These are double drums with strings that holds small 'beads'. When one rotates the drum back and forth, the beads tap on the drum.

Swords, like the one of Buddha Manjushri, symbolise the ability to cut through delusions or obstacles, and as such they represent wisdom, knowledge, or the protection given by Buddhist doctrine. On the image at the left, it stands on the text of the Heart Sutra, a crucial teaching on emptiness in the Mahayana tradition, cutting through ignorance.

The Kartika (Skt.) or curved knife symbolises the cutting of conventional wisdom by the ultimate insight into emptiness. Is is usually present as a pair, together with the skullcup, filled with wisdom nectar. On a more simple level, the skull is a reminder of (our) impermanence. Between the knife and the handle is a makara-head, a mythical monster.

The Khatvanga (Skt.) could be called a magic wand or magicians' stick and represents the 'magic powers' or siddhis (Skt.) of an accomplished tantric practitioner.

"The shaft of the khatvangha has eight sides which represent the Noble Eightfold path (the fourth Noble Truth) and the eight classes of protectors. At the end of the shaft is a dorje representing totality and completion. Along the shaft of the khatvangha are crossed dorjes, a gterbum and three heads. The crossed dorjes are symbolic of the indestructibility of beginningless wisdom mind. The gterbum is symbolic of wealth and enrichment. The three heads-one freshly severed, one rotting and one a skull-are the symbols of the three spheres of being, chö-ku, long-ku and trül-ku [Nirmanakaya, the middle one represents the Sambhogakaya, and the top one is a skull, representing the Dharmakaya] which are unified by the shaft of the khatvangha demonstrating their inseparability. Streamers of the colours of the five elements hang from the khatvangha, as well as a bell and dorje which represent emptiness and form. At the top of the khatvangha are the three prongs which pierce the fabric of attraction, aversion and indifference. Hanging from the prongs are two pairs of rings. These signify the four philosophical extremes that are denied by Dharma: eternalism and nihilism, monism and dualism. Finally, the khatvangha is surmounted by wisdom fire—the fire that burns self-protection, justification and referentiality."

The top of the kathvanga can be formed by a vajra or a trident (often depicted with flames around it)

The hook (ankusa in Skt.) is a tool which was originally used to ride elephants. In Buddhism, it is considered a hook with which one can draw people in doubt to the teachings.

A wrathful implement like the Phurba (Tib.) or ritual dagger is used to subdue and subsequently enlist negative, uncontrolled in a wrathful manner. The cutting end has three sides, rather than a normal dagger. Above the cutting end is a makara-head (a mythical monster) and at the top is the handle, made of three heads of wrathful protectors. Phurbas refer to the Buddhist theme of self-control, and to the tantric transformation of negative powers on the path to enlightenment.

The Tibetan Bumpa is a ritual vase which represents the palace of the deities. It is used as a vessel of purification, to bestow blessings and confer empowerments. There are also Treasure Vases, which contain special substances and are buried or hidden for various purposes.

·Mallets or Hammers (Skt. mudgara) and a Club (Skt. gada) symbolise crushing strength or power.

Bow and Arrow refer to single pointed concentration to achieve the goal of liberation.

The Trident is a piercing weapon, its, three points also carrying connotations of the power of the three jewels.

An Arrow with ribbons around it can symbolize longevity and prosperity.

The Lasso relates to the constraint of negative forces.

A Trident symbolizes the attainment of the three Kayas (or Buddha bodies).

Mudras—Sacred Gestures: To make the tantric experience complete, actions of one's body, speech and mind are to be transformed. A typical example of the actions of the body is the practice of mudras. These are movements and positions of the hands which have profound symbolic meaning. One uses mudras to symbolise for example the various offerings, but they also convey general meanings like, *e.g.* teaching or meditation.

The image on the left is a virtual image of the planned Maitreya statue in Bodhgaya. It happens to be an unconventional posture to depict Maitreya: his right hand on the knee signifies giving refuge and loving compassion to all beings; the left hand at his heart is in the teaching (Dharmachakra) mudra: the thumb and index finger are pressed together to symbolise the united practice of method and wisdom, and the three remaining fingers are raised to symbolise the Three Jewels of Refuge-Buddha, Dharma and Sangha. To the right is an image of the teaching mudra when performed with both hands. The image on the right represents a traditional depiction of Shakyamuni Buddha, just after his enlightenment.

His right hand touches the earth to symbolise it has witnessed his enlightenment (bhumisparsha mudra), and his left hand is in the meditation mudra. Sometimes the meditation mudra is performed with both hands on top of each other in the lap as the image below shows.

Mandalas—Sacred Circles: In Tibetan Buddhism, Mandalas come in two varieties; it can represent the universe, as it is used in the Mandala Offering Ritual, where one symbolically offers the entire universe. For this, several rings can be placed on top of each other filled with rice and precious objects. During the offering one recites mandala offering prayers. In the centre of the mandala is Mount Meru, the central axis in the Buddhist (and Jaina) cosmos.

The best-known mandalas are part of the world of Tantra; they represent the "3D Palace" of a specific meditation-Buddha or deity. In the Tibetan tradition, they come as thangkas (scroll-paintings), wall paintings, sand-drawings and 3D models of, *e.g.* wood or metal.

A mandala can be "read" and studied like a text and most important, be used for tantric meditation. The purpose of a mandala is to acquaint the student with the Tantra, and thus allowing the student to identify with the deity and its sacred surroundings as the mandala.

The mandala of Kalachakra like the one on the left, symbolises the entire universe in terms of planets and time cycles, as well as aspects of our body and mind, and even the practice. Deities and other images within the mandala represent for example the sense organs, the elements and mental aspects, all in a purified state. Every implement held by each deity has again its own meaning.

A sand mandala represents a full 3 dimensional palace, often with more than one storey. For example, the image on the right is from an actual 3D mandala of Zhi-Khro (the six bardos) with 2 stories. These mandalas are used in many ways; as a focal point of intense meditation, in the practice of actually making one, and in their destruction (like sand mandalas after an initiation is completed) to teach impermanence. In the practice of Kalachakra, one strives at visualising the complete mandala, including its hundreds of deities in perfect detail within the size of a tiny drop. This should give an idea of the level of concentration required for transforming oneself into a Buddha.... Creating a Kalachakra mandala of some 2 metres diameter, traditionally takes 6 days, employing as many as 16 monks. For a nice description of the Chenresig mandala.

Symbolism of Some Deities

In most of the mandalas of tantric Buddhaforms, the 5 Dhyani Buddhas are mentioned; some of the symbolism related to them are found on the 5 Dhyani Buddha page.

Each image of a tantric deity, including Shakyamuni Buddha himself is filled with symbolism related to their specific qualities. In Tibetan Buddhist monasteries, one can often also find very aggressive-looking images, they are called the Protectors or Guardians. Many-but not all-of them are manifestations of fully enlightened Buddhas, but they appear in this wrathful form to 'wake up' the meditator and for example direct this powerful energy towards ones own problematic emotions like anger, attachment and ignorance.

From The Crystal and the Way of Light: Sutra, Tantra and Dzogchen, Teachings of Chogyal Namkhai Norbu, compiled and edited by John Shane:

> "*Guardians of the teachings*
>
> *There are eight principal classes of Guardians each with many subdivisions. Some are highly realized beings, others not realized at all. Every place-every continent, country, city, mountain, river, lake or forest—has its particular dominant energy, or Guardian, as have every year, hour and even minute: these are not highly evolved energies. The various teachings all have energies which have special relationships with them: these are more realized Guardians. These energies are iconographically portrayed as they were perceived when they manifested to masters who had contact with them, and their awesome power is represented by their terrifyingly*

ferocious forms, their many arms and heads, and their ornaments of the charnel ground. As with all the figures in tantric iconography, it is not correct to interpret the figures of the guardians as merely symbolic, as some Western writers have been tempted to do. Though the iconographic forms have been shaped by the perceptions and culture of those who saw the original manifestation and by the development of tradition, actual beings are represented."

Various other Symbols: The scorpion refers to negative or harmful action (also menace or threat). It refers to transformation, simply said, 'if the scorpion can be transformed then anyone, and anything can be transformed.' Everyone and everything-is nothing other than the energy of the non-dual state—and therefore the power of every facet of existence can be harnessed through pure vision as a means of attainment and compassionate activity.

Tibetan Buddhist Symbols

Tibet is a vast country, spread out between India and the Himalayas in the South, and China in the east. The traditional nomadic culture was greatly influenced by the introduction of Buddhism from India, to an extent that nearly all the cultural achievements of Tibet are related to the Buddhist religion. Because of the predominance of Tantra with its rich tradition of symbolism, it is no surprise that symbols and symbolic artifacts of all sorts are found in Tibet. Some of the symbols however originated in Tibet, or were given a specific meaning within the local culture.

The Kalachakra 'Tenfold Powerful' Logo: This symbol can be found nearly everywhere where Tibetan Buddhism is present, in various forms. It represents the teachings of the Kalachakra Tantra, one of the most complex tantric systems. This symbol was developed in Tibet and is a schematic representation of letters in the Lantsa script. The symbolism behind this logo is vast and has explanations referring to the outer world, the human body at its gross and subtle levels, and the practice of Kalachakra. A more extensive explanation can be found on this page of the International Kalachakra Network.

Prayer Flags, Mani Wheels and Mani Stones: Typically found in Tibetan areas are the many Prayer Flags that decorate monasteries, houses and even mountain passes. It is believed that the special blessing power of the mantras printed on them is spread all over the world by the wind. The tradition stems from pre-Buddhist traditions in Tibet, just as the so-called Wind horse which is often found on them. Traditionally, prayer flags have a combination of mantras, special prayers and auspicious symbols printed on them.

Prayer flags are not only strung on ropes to hang between two poles, they can also be hung like a flag from one pole.

Usually, the flags come in five different colours: blue corresponds with the sky, white with clouds, red with fire, green with water and yellow with earth. (This is different from the usual general symbolism of colours, which is related to the elements: black/blue for air, red for fire, white for water and yellow for earth).

The Wind Horse (Tib. Lung-ta), a mythical Tibetan creature from pre-Buddhist times, combines the speed of the wind and the strength of the horse to carry prayers from earth to the heavens. Not surprisingly in a country where the horse was used by the traditional nomads of Tibet. It is associated with success and the space element. The Wind Horse carrying the "Wish Fulfilling Jewel of Enlightenment" is the most prevalent symbol used on prayer flags. It represents good fortune; the uplifting life force energies and opportunities that make things go well. When one's lung-ta is low obstacles constantly arise. When lung-ta is high good opportunities abound. Raising Wind Horse prayer flags is one of the best ways to raise one's lung-ta energy.

"The traditional Wind Horse Prayer Flags are ancient designs. In the centre of the flag on the right, is the Wind Horse, the uplifting energy that carries good fortune to all beings. In the corners are the "Four Dignities": the Garuda (wisdom), the Dragon (gentle power) the Snow Lion (fearless joy), and the Tiger (confidence). The Eight Auspicious Symbols are depicted around the perimeter. Various prayers and mantras are included in the text.

At the top centre are the three main Bodhisattvas: Avalokiteshvara (Compassion), Manjushri (Wisdom), and Vajrapani (Power). At the bottom centre are the figures called "the union of enemies" representing friendship. The two swastikas are symbols of eternal life. There are many prayers for health, prosperity and good fortune."

Prayer Wheels are another typical Tibetan phenomena. They are usually crammed with mantras, and similar to prayer flags, it is taught that the power of the mantras will spread when the wheel is turned. The more mantras, the better. Prayer wheels come in many variations; from handheld, like the one on the left, to huge ones that require serious muscle power to set in motion. They can be moved by hands, water or wind-with this cool tool, even your hard disk can function as a prayer wheel. They are also called Mani-wheels, as the wheels usually contain the mantra OM MANI PADME HUM (the letters below show this famous mantra in Tibetan), which is the mantra of Chenresig (Avalokiteshvara in Sanskrit). A thorough explanation of prayer wheels can be found here, and teachings of Lama Zopa Rinpoche on the benefit of prayer wheels.

The following story describes the origin of the prayer wheel:

"Arya Chenrezig predicted to Master Ludrup Nyingpo, "In the palace of the land of Naga is the Naga King Bodhisattva, who is the owner of a profound wheel of Dharma. By hearing, seeing, touching or thinking of this wheel, one can swiftly attain liberation from the suffering of the three lower rebirths. If you go and fetch this wheel, the benefits to sentient beings will be enormous."

Consequently, Master Ludrup visited the land of Naga and said to Naga King Bodhisattva, "Oh, Naga King Bodhisattva, please pay attention to me. I have come here because Arya Cherezig prophesized that the benefits to sentient beings will be enormous if I beg from you your profound wheel of Dharma, which can liberate beings from all types of sufferings of lower rebirths just by hearing, seeing, touching or thinking of it. Kindly give it to me."

Naga King Bodhisattva replied, "This wheel of Dharma, which has the quality of quickly liberating all transmigrators from the great suffering of the three lower rebirths merely by hearing, seeing, touching or thinking of it, was kindly given to us in the past by the Buddha Mar Mezed, and has given nagas much happiness, through it many have been led to the grounds and paths of Buddhahood. This Dharma Wheel is the wheel of the mantra Om Mani Padme Hung, the essence mantra Arya Chenrezig received from the Buddhas upon request, and which represents the essence of all the qualities of body, speech, mind and actions of the Buddhas. I shall give this wheel to you. You must place it on or in earth, water, fire or wind and use it for the sake of Dharma and living beings."

The wheel was passed on to Master Nagarjuna together with its instructions for use. Master Nagarjuna brought it to India and later passed it on to the Lion-Faced Dakini. From her the lineage passed through Tilopa, Naropa, Marpa and Milarepa to Dakpo Lha Je, then to Khampa Yu Se and so on."

Another, very Tibetan, phenomena is to cut mantras into stones, and pile them up to form entire walls. As the mantra carved onto the stones is again often the Chenresig mantra "OM MANI PADME HUM", they are also known as Mani-stones, often combined into Mani-walls.

Khata Offering: It is customary to offer white scarves (Tibetan: khatag) at this time, or at the very end, but this is not strictly necessary. If a khata is offered (held as in the picture) then it should be placed in front of the lama or at his side. Often, it will be returned to you (with the help of the assisting monk) as a blessing, but not always. They may be available for purchase before the ritual, but if you do not have one, you can offer a smile and a bow.

There is often a natural inclination to make offerings to the lama out of recognition for his or her kindness. Plan to bring an envelope (or two, if you would like to donate something to the translator or other assistant) into which you can put your contributions. It is customary to make these donations at the end of the wang. You can place this on the lama's lectern or desk with the katha.

Though the value of an empowerment cannot be measured in dollars, that does not mean that money is unsuitable as an offering. Flowers and fruit are suitable as offerings to Buddhas and Bodhisattvas; lamas who are residents at dharma centres, who help build monasteries and schools, and who travel widely need money. Each person should offer what he or she feels is appropriate.

In Mongolia, the people often offer blue khatas.

The Four Dignities: The Four Dignities are mythical animals which represent various aspects of the Bodhisattva attitude, like strength, protection and cheerfulness.

Dragon: The Dragon thunders in the sky with the sound of compassion that awakens us from delusion and increases what we can know through hearing. Dragons have the power of complete communication. Just as we do not see sound, we do not see dragons—at least not usually. Displaying a dragon banner is said to protect one from slander and enhances one's reputation.

Associations: Main quality is power, dominance over the sea, and the water element.

Tiger: The Tiger abides in the South, symbolizing unconditional confidence, disciplined awareness, kindness and modesty. It is relaxed yet energized; resting in a gentle state of being that has a natural sense of satisfaction and fulfilment, referring to the state of enlightenment.

Associations: Main quality is confidence, dominance over forest, and the air element.

Snow Lion: The Snow Lion resides in the East and represents unconditional cheerfulness, a mind free of doubt, clear and precise. It has a beauty and dignity resulting from a body and mind that are synchronized. The Snow Lion has a youthful, vibrant energy of goodness and a natural sense of delight. Sometimes the throne of a Buddha is depicted with eight Snowlions on it, in this case, they represent the 8 main Bodhisattva-disciples of Buddha Shakyamuni, the historical Buddha.

Associations: Main quality is fearlessness, dominance over mountains, and the earth element.

Garuda: The Garuda is daring and fearless and abides in the north. With great strength and power it soars beyond without holding back. It symbolizes freedom from hopes and fears, the vast mind without reference point. It is a powerful antidote to the negative influences of Nagas (spirits) which can cause disease and all kinds of harm.

Associations: Main quality is wisdom, dominance over the sky, and the fire element.

The Three Symbols of Victory Against Disharmony: These symbols can easily be confused when not looked at carefully; for example, this Snowlion has a Garuda head....

The pairing of the six traditional enemies-garuda and snowlion, otter and fish, crocodile (sea-monster) and seasnail-creates the three mystical animals, symbols of universal love.

"The Union of Opposites (mithun gyulgyal) is an interesting group of symbols. These mythological beings are joined rival pairs of animals created to symbolize harmony. A snow lion and a garuda, normally mortal enemies, were combined to form an animial with a snow lion's body and a garuda's head and wings. Likewise a fish was put together with an otter and a crocodile-like chu-srin was married to a conch shell. These composed creatures are often put on Victory Banners for the reconciliation of disharmony and disagreement."

Mount Kailash: Tibetans consider a pilgrimage journey around Mt. Kailash very important, some people even manage this high-altitude, difficult trek by making prostrations. It is even said that if one does 108 circumambulations, one will reach enlightenment. Quite a number of tales are connected to the mountain, linked with the greatest saints of Tibetan Buddhism, Guru Padmasambhava and Milarepa, and is the abode of the deity Chakrasamvara.

"There is no place more powerful for practice, more blessed, or more marvellous than this; may all pilgrims and practitioners be welcome!"

Milarepa: In Bon-Tibet's pre-Buddhist religion the mountain is called Tise and is believed to be the seat of the Sky Goddess Sipaimen (Hindus identify it with the abode of Shiva. Jains call the mountain Astapada and believe it to be the place where Rishaba, the first of the twenty-four Tirthankaras attained liberation).

Karmapa Dream Flag: The 16th Gyalwa Karmapa, well known for his visions and prophesies, designed this flag from a vision that came to him in a dream. He called it "Namkhyen Gyaldar (Victorious Flag of Buddha's Wisdom)." He proclaimed, "Wherever this banner is flown the Dharma will flourish."

According to the outer meaning the blue represents the sky or heaven. The yellow symbolizes the earth. The wave symbolizes the Buddhadharma penetrating heaven and earth.

According to the inner meaning, the blue represents vision and spiritual insight and the yellow symbolizes our experience of the everyday world. The symmetry of the wave pattern shows the interdependence of the absolute and relative levels of reality.

According to the secret meaning, the blue symbolizes emptiness-wisdom and the yellow represents compassionate action. The wave is Mahamudra: the union of compassion and wisdom—the ultimate realization of one's true nature.

Musical Instruments in Tibetan Buddhism: To make the offering of sounds, the Tibetans use a whole range of musical instruments, from bells, cymbals, drums to wind instruments. Especially some of the wind instruments may produce a very strange sound for the Western ear, like the Long Horns which sound almost like a toneless blare to the Thighbone Trumpets with a sometimes piercing sound. To judge for yourself, do visit this page of the Nuns of Khachoe Ghakyil, which include descriptions and even soundbytes of the instruments.

Tibetan Thankas: A specific artform that has developed into perfection in Tibet are the scroll-paintings or Thankas. On one hand the often nomadic people liked to have images of deities and other subjects, but transportation of conventional flat paintings was problematic. That is why the art of making paintings that can withstand rolling up was developed into perfection. Traditionally, very specific recipies are used for making the paint (mostly made of natural minerals) and preparation of the canvas.

Chakras and the 64 Yoginis

As the first of the siddhas, the Tibetan sources mention Luhi-pa (Luyi-pada) who is most probably identical with Matsyendranatha, the teacher of the famous Gorakshanatha... The natha-siddhas...deserve to be singled out for separate treatment by virtue of their enormous influence on the development of Yoga.

There are more systems of chakras in the tantrik tradition than most people realise. One of the most ancient appears in chapter nine of the Kaulajnana Nirnaya, attributed to Siddha Matsyendranath. Said to be the progenitor of the Kaula school of Tantra, this chakra system is eightfold. Each of the lotuses has eight petals and these represent the 64 yoginis, famed in Kamarupa.

"The letter ksa is in the Brahmarandhra, la in the forehead, ha between the eyes, sa in the mouth, sa in the throat, sha in the heart, va in the navel, and ha in the genitals. Dear One, listen to the collective meditation in these chakra places.

"The first of the eight is a celestial eight petal lotus of pure crystal effulgence, a heavenly fire or flame, completely without stain, free from all duality, from the fires of cruelty, devoid even of nothingness. One should avoid perturbation and practise equipoise (Sama) consciousness. One my see even beyond the future, one with the endless circle of time, having knowledge of what is spoken at a distance, able to both grant boons and to paralyse, with the ability to seize and subjugate pashus, or to destroy or to kill them, mortal, always truthful, and in equipoise, becoming eloquent, a Siddha, and able to do anything.

"One should meditate on the second as having eight petals, as bright as a beautiful pure flame. One conquers death and becomes the cause of great excitation. Meditating one-pointedly on this multifold form one becomes an emperor amongst kings, doing whatsoever one wills, and may destroy all in the three worlds, whether animate or inanimate-like Kruddha, the cause of both creation and dissolution, always engaged in love.

"The third great chakra is the deliverer from the nine tattvas. After meditating on the Guru in that chakra for six months, one my achieve whatever is wished for, destroy old age, be able to see at a distance, able to obstruct, even if one hundred yojanas distant. In a lonely place one should meditate on that multifold form.

"The fourth chakra is the cause of peace, increasing happiness and pleasure. By always meditating on it, afterwards one becomes immortal, eloquent, victorious over death and disease. Practising it daily, one causes death to flee. By continual practice for a period of sixteen seasons on this great chakra, one becomes able to destroy in a day and a night.

"O Surasundari, the fifth great lotus has eight petals. One should always meditate on it as being of a smoky colour, then one my shake the three worlds. It causes eloquence and avoidance of untimely death, and allows one to both enslave and paralyse.

"If one should meditate devotedly on the sixth royal chakra, fiery, with a pericarp of eight petals, as bright as liquid gold, the cause of Iccha Siddhi, it enables one to see into the future, giving Anima and the other seven siddhis—of this there is no doubt.

"Dear Mahadevi, the seventh is as bright as the full moon, auspicious, within the body, bestowing both enjoyment and liberation, destroying fever and death, enabling one to enter into another's body. what may not be done? One becomes the best within a circle of people. "Pretty-Eyed One, the eighth great chakra, adorned with eight petals, gives Dharma, Artha, Kaama and Moksha."

Chakras According to Goraksanath: Whatsoever difference there has been, or may be, as to forms and methods, whether in Upasana or Yoga, yet all Indian worshippers of the ancient type seek a common end in unity with Light of

Consciousness, which is beyond the regions of Sun, Moon and Fire-The Serpent Power, Arthur Avalon, p.287

The following article, called "System of Chakras according to Gorakshanatha," by the esteemed tantrik scholar Gopinath Kaviraj, first appeared in the Princess of Wales Saraswati Bhavan Series, Vol II, 1923, and is consequently out of print. Words which were in Sanskrit in the original article have been converted to the itrans format.

The System of Chakras According to Goraksanath: The system of mystic culture introduced by Goraksa Natha does not seem to have spread widely through the educated classes, so that although about a millennium has passed since this great man appeared his teachings have remained till to day almost a sealed book to many. The Goraksa Sataka and Goraksa Paddhati are two of the few Sanskrit works published which profess to give an exposition of this teacher's instructions. The Hathayogapradipika also belongs to this school, but it deserves to be supplemented.

We propose to deal in a separate paper, on the basis of Mss. and printed books (in Hindi and Sanskrit), with the entire system of mystic culture associated with Goraksa Natha, both historically and philosophically. Here we must confine ourselves to the system of Chakras or intra-organic centres of spiritual energy recognised by Goraksanatha. Our description will be based on a Ms of Vairata Purana and on an old coloured chart (obtained accidentally from a local gentleman interested in this panth) representing: the human body containing the spinal column with the various centres painted and duly located. It will be found that this arrangement differs widely from the current notions both of the Hathayoga school and of the Tantras.

First of all in the perineum we have the Adhara chakra (coloured red) presided over by Ganesa natha with his two Powers, *viz.* Siddhi and Buddhi. This is identical with the well known Muladhara of the Tantras. But the next centre, called Mahapadma chakra, controlled by Nila natha is unknown else where, The third, the Svadhisthana chakra (coloured yellow), is in the genital region and has Brahma for its Deity and Savitri for the Power. Between this and the Manipura there are three distinct centres, *viz.* Saddala (called also Susumna chakra), Garbha (in the Garbha sthana) and Kundalini (in the region adjoining the waist and presided over by Fire). Besides bare names and vague localisation we do not find statement for any further detail about these psychic vehicles. The Manipura is situated in the navel and has Vishnu for its Devata. Above this is supposed to exist the so-called Linga chakra, of which, again, no particulars are given. Higher still, in the pericarp of the Anahata is the seat of mind—Manas.

The Anahata itself is in the heart and looks like a lotus with 12 petals, emitting a white radiance around. The presiding god of the chakra is named Mahadeva (Rudra natha, in the Ms) and the Power is Uma. The rsi is called Hiranyagarbha. This corresponds to the causal body, dreamless sleep, Pasyanti Vak and Sama Veda.

The next higher Chakra is of course Visuddha, in the throat. It is a sixteen-petalled lotus, with smoky colour, presided over by Jiva and Adya Shakti, The Rsi is Virat. It corresponds to the causal body, dreamless sleep, Paravak, Atharvana Veda, Jalandhara Bandha and Sayujya Mukti.

The Prana chakra which is a thirty-two-petalled lotus of bright hue (ududyotavarnaprabha) and is controlled by Prananatha and Parama Shakti, is seated near the region of the throat (galasthana). It forms the 10th aperture of the human body. Of the four chakras above Visuddha and below Ajna, the second one is Avala chakra, furnished with 32 lobes shining 'like the rising sun (arunodyotaprabha), presided by Fire. The exact chapter of this Chakra is not mentioned. From what is said it appears that it is seated where the three granthis, *viz.* Brahma, Vishnu and Rudra, unite (trigranthisth Anagzg), and is very intimately connected with Kalachakra and Yoginichakra. The Chivuka chakra is somewhere in the facial region, apparently near the chin, and is formed like a sun-like lotus of 34 lobes, presided by Prana and Saraswati. All the devas have their seats within the lotus. Its Rsi is named Krodha. All languages, indeed human speech itself, are supposed to have their origin here. The Balavan chakra is just below the Ajna, in the nasal region, and looks like a three-petalled lotus of red, white and dark colour. This place is described as Tribeni, being the confluence of the three streams of Ganga, Yamuna and Saraswati represented in the body by the three nadis, *viz.* Ida, Pingala and Susumna. The presiding God of this Chakra is Pranava and the Power Susumna. The statement that this place is associated with the three matras of Pranava (*viz.* A, U & M) becomes thus intelligible. The name of its Rsi is given as Mahahankara.

The famous Ajna chakra (called also Ani chakra in the Ms.), which is in the centre of the space between the two brows, is a diamond like (manikyavarnaprabha) lotus of two petals, presided by Hamsa Devata, and Susumna Shakti. It corresponds to the Vijnana state and Anupama Vak, and to the half matra of the Pranava.

The Karnamula chakra, within the auricular region, is a 36-lobed lotus of mixed colour (dark and yellow.) The presiding God and Power are Nada and Sruti respectively.

It is the seat of the 36 matrkas.

The Tribeni chakra, above the brows, is a 26-lobed circle with Akasa as its rsi. This is the real Tribeni, but how this place is connected with the Balavan chakra lower down is no-where stated.

The Chandra chakra is in the forehead and consists of 32 lobes with a colour between white and red. It is presided by the Moon and Amrta Shakti. The Rsi is Manas (mind) with its sixteen kalas. It is said that the sun goes to this lunar mansion to drink nectar.

This centre is very closely related to another chakra—the Amrta chakra, almost in the same region, probably a little upwards. Its Devata and Shakti are identical with those of the preceding chakra, but the Rsi is Atma rather than Manas.

It is from here that nectar is constantly flowing, This place is described as the abode of the Gayatri named Kamadhenu (lit. 'wish-giving cow'), figured like a milch cow with four teats, *viz.* Ambika, Lambika, Ghantika and Talika. It has a human face with intoxicant looks (madanetra), a peacock's tail, a horse's neck (griva), an elephant's tusk (tunda), a tiger's arms (hastashardula), a cow's horns, and wings consisting of Lila Brahma and Hamsa. It is a stranger figure. It is from the udderof this 'cow' that nectar is perpetually streaming down. The Khechari, Viparitakarani and other mudras are some of the devices intended to check the downward flow of this blissful current. The Yogin who has obtained access to the chakra and abides here becomes immortal and free from the effects of Time.

Next is the Brahmadvara Chakra, located above the forehead and shining with its 100 petals like the many coloured rainbow; and beyond this is the seat of the Akula Kundalini—a lotus of 600 petals bright like the newly risen sun.

On crossing this one comes up to the Brahmarandhra in the cranium (murdhasthana), with its multi-coloured 1,000 petals. This is the so called Sahasrara of the mystic literature-the Aim and End of all spiritual progress. It is here that the Guru and the Chaitanya Shakti reside.

One would naturally expect that the series of chakras would come to an end here. But the picture on which the above account is based, mentions 6 chakras more beyond the Sahasrara, *viz.* (a) Urdharandhra, (b). Bhramaraguha, (c) Akunthapitha Punyagara, (d) Kolhata, (e) Vajradanda and (f) Niradhara Paramajyotischakra. The Urdharandhra, called Talu chakra in the Ms, is seated in the Talima and is a 64-lobed chakra, presided by Goraksanatha and Siddhanta Shakti. This statement is interesting as it gives us an idea of what the followers of the path thought in connection with the founder of their school.

The Bhramaraguha, also called Alekha, *i.e.* Alaksya chakra (called Brahma Chakra in the Ms), is the place, where samadhi-yoga ensues, and prana and manas cease to work. The lotus is described as furnished with ten millions of lobes and wonderfully brilliant. The presiding God of this centre is Alaksyanatha, the Shakti called Maya (= Mahamaya?; Akula in the Ms.) and the Rsi MahaVishnu.

The higher Chakra with an equal number of lobes has Akalanatha as Devata, Akalesvari as Shakti and as Akala as Rsi.

The Kolhata Chakra is in the Sikhamandala and corresponds to the Vaikuntha of the Vaishnavas and Kailash of the Shaivas. Both the Devata and the Shakti are named Ananta. The Ms. calls this region a road to the Highest Void (paramashunyamarga).

The description of Vajradanda is not very clear. It is said to be, as I understand it to be, in the form of a column, vast (mahavishala), lustrous (tejahpunjaprabha) and long (dirgha).

The final Chakra is in the Niralambasthana, with an infinite number of lobes, colours, matrkas, devas and worlds. This is the Highest Seat of the Gurudeva.

Beyond this is a series of 20 voids of which nothing is said. The Ms. observes that Final Liberation takes place in the Great Void (paramashunyasthana) above 21 Brahmandas. Transcending the great Void the Yogin becomes eternally free from 'coming and going', *i.e.* the wheel of birth and death: sa cha yogi tishthati yuge yuge jyotih sametya.

We have tried to reproduce above with as much fidelity to the understood meaning of the chart as possible the account of the Gorakhpanthis concermng the arrangement and function of the various chakras. But as the chart was on an old canvas with the figures rendered indistinct by time and the notes appended generally vague, illegible and sometimes totally unintelligible it is quite likely that mistakes and in some places even confusion have been left unnoticed. It is sure nevertheless that the general presentation of the scheme is unimpeachable.

Taking it as we find it, there is no doubt that the ideas of this school differed in many points from those of the Tantras on the same subject. The question of the total number of chakras may be left aside for the present. For we are aware that there are several hidden (gupta) chakras which are not ordinarily counted; and very often a certain number of chakras, considered minor from one's own point of view, is supposed to form a connected whole. There are other reasons also which would explain the difference of the number in different systems. The actual description of a particular chakra is more important to a student for purposes of comparative studies. But even here there are certain fatal limitations. For instance, the same chakra may not look exactly alike to different sadhakas. The personal sankalpa of the aspirant, either conscious or sub-conscious, and that of his Guru go a great way towards determining the feature of the Vision. The reality observed is, in a certain sense, only ideal.

Entering into detail we may observe that the Sahasrara is not supposed to form the Highest Limit:-there are stages beyond. With this we may compare the accounts of the Radhaswami sect who also hold more or less the same views. The Manasachakra as described here embodies 32 lobes, while elsewhere it is said to be 6-petalled (Dr. B. N. Seal, The Positive Sciences of the Ancient Hindus, p, 221; Avalon, The Serpent Power,. p. 146).

Regarding akulakundalini it may be said that the Tantrists locate it within the Moon of Consciousness which forms the pericarp of the downward facing Sahasrara and is situated in the transcendant heaven (para vyoma)—a technical term for a part of the cerebral region. The contact of kula with this akula is the immediate cause of the flow of nectar (sudhasrava). While Goraksanatha holds that the nectar flows from the Amrta chakra above the Moon, Bhaskara considers that it flows from the Akula which is within the Moon.

The name of Bhramara Guha is to be found mentioned in the literature connected with the names of Kabir, Radhaswami, etc., but nowhere is its function clearly stated. The Sutasamhita and Bodhasara use the term vaguely in the sense of Brahmarandhra. This so-called cave is in reality a hole or rather a hollow which appears to view when one gazes into the centre of the 'Kutastha.' The entrance to this hollow is brilliantly dark, but it is surrounded by a luminous ring of rays.

The powers of obscuration (Avarana) and dissipation (vixepa) are said to guard this entrance, so that they try to screen up and throw off the gazer. It is only when the power of introvision is produced after the generation of pure magnetism within the body through strict continence, pure food, &c. and through the cultivation of the moral virtues of selflessness, forbearance, &c. and through the action of prana that one can expect to gaze at this hollow without fear of Avarana and vixepa. At this state mere gazing makes the mouth of the cave wide open, and every tattva is illuminated.

In the Chart the Prana chakra is described as the tenth avenue of the human body. This aperture is usually closed in men, so that the body is as a rule likened to a "city with nine gates". (Cf. navadvare pure dehi-Gita) only. But a steady process of psychic discipline helps to open this avenue, through which the Jiva of the Kramamukta-upasaka passes away along the ray of the sun into the Solar Region, called also Brahmaloka, and thence with the dawn of Knowledge is absorbed in Brahman.

The medulla oblongata (bhastaka granthi) above the Visuddha Chakra, is one of the sites where the three nadis are united. From here the Susumna enters into the skull, and the other two nadis, *viz.* Ida and Pingala, pass along the right and left sides of the forehead and meet together and are joined with the Susumna between the two eyebrows. From there, the Ida goes to the left nostril and the Pingala to the right.

From the medulla the Susumna is bifurcated: one line passes below the brain and in a rather oblique course comes to the eyebrow whence with a slightly upward bend pierces the pericarp of the Ajna and unites with Ida and Pingala.

Then it comes out, and running straight up crosses a very subtle hole within the interior of the central region of the forehead and hanging down to some distance takes a curve and goes right up, penetrating the Sahasrara and entering the Brahmarandhra. Another line goes up direct from the medulla, and through the interior of the skull extends to the Sikhara. With a slight curve it enters the Brahmarandhra. The mouth of this line of the Susumna which is in the Brahma randhra remains usually closed, while that of the first line is open. Consequently the hollows of the two lines are not in union.

While passing away from the body the yogin gets the closed mouth of the Susumna opened, on which the two holes mentioned above become unified. This is what is usually known by the name of the "10th avenue". In the Amaraugha Sasana, however, the 10th aperture is identified with the mouth of the

Sankhini, which is a hollow behind the front tooth (rajadanta) and the Kankalamalini Tantra locates Brahmarandhra just below the Sankhini.

The Gorakhbodh

Gorakh: O Lord (Svami), you are the Master Teacher (Satguru Gosain), and I am but a disciple: may I put a question, which you would kindly reply to and resent not? To start with, what ideal (lacch) should the disciple put before him? Do tell me for you are the true Teacher.

Macchendra: Let the unattached (awadhu, avadhuta) live at the monastery (hat) or be on the road, resting in the shadow of the trees; he should renounce desire, cruelty, greed, delusion, and the illusion of Samsar (Kaama, Krodha, Lobha, Moha and Samsar ki Maya); he should hold converse (gosht) with himself and contemplate the Endless (Ananta); he should sleep little and eat little. In the beginning the disciple should live thus. Thus speaks Macchendra.

G: What should he see, what contemplate, and what treat as the essence (sar); with what should he shave his head and with the knowledge of what should he try to cross (the ocean of Samsar)?

M: He should see himself, contemplate the Endless (Ananta), and fix upon Reality as the essence; he should shave his head with (or after receiving) the word of the teacher (Guru ka Shabda), and should cross over with the aid of Divine knowledge (Brahma Gyana).

G: What is the teaching (upadesh) of the Guru's order or doctrine (Ades)? Where does the void (Sunya) reside? Who is the Guru of the word (Shabda)?

M: The most wonderful (anupam) is the teaching of the Guru (Ades); the void (Sunya) resides within us and Realisation (parcha or parichaya) is the Guru of the word (shabda).

G: What is the form (rupa) of the mind (mana)? What is the appearance (akar) of the vital breath (pavana)? What is the direction (disa) of the ten and through which door can the control be effected?

M: The void (sunya) is the fore of the mind; the appearance of vital breath (pavan) is invisibility (nirakar); the direction of the ten is unsayable (alekh) and control lies through the tenth door.

G: Which is the root (mula) and which the branch (bela)? Who is the Guru and who the disciple; with what essence (tatt) can one go about alone?

M: Mind is the root and vital breath is the branch; the word (shabda) is the Guru and attention (surat or surta) is the disciple. With the essence called deliverance (nirbana tat) Gorakhnath wanders about, himself in himself.

G: What is the seed (biraja) and what the field (khetra)? What is direct hearing (satvan)? What is true vision? What is Joga and what is the method (Jugti)? What is liberation (mocch)? And what is salvation (mukti)?

M: The word (Mantra) is the seed; perception (mati) is the womb or land; and attention (surti) is direct hearing, and discrimination (nirti) is true vision; the ocean (Uram) is Joga and the earth (Dhuram) is the method; light (joti) is liberation and the refulgence (Juala) is salvation.

G: Which is the tree without a trunk, and which is the parrot without wings? Which is the dam (palu) without a shore (tir), and who died without death (kal)?

M: Vital breath (pavana) is the tree without a trunk; mind is the parrot (sua) without wings; constancy (dhiraj) is the dam without a shore; sleep is dying without death.

G: In what house (ghar) is moon (chanda) and in what is the sun (sur)? In what house does Time play music (Tur, a musical instrument)? Where do the five elements (tat) have equipoise (sam rahai)?

M: The moon in the mind; the sun in the vital breath; in the void (sunya) Time plays on the musical instrument (tura) and in the house of knowledge the five elements reside in equipoise (sam).

G: What is the New Moon (amavas) and what manifests (pariba)? Which or where is the great elixir (maha rasa) and whereto with it do we mount? At what place does the mind reside in the state of self-transcendence (unmani)?

M: The sun (ravi) should be treated as the darkest night; the moon should be made manifest; the great elixir of the lower (ardh) should be taken to the upper (urdh); in the heaven within us (gagan) the mind resides in self-transcendence.

G: What destroys the bad word (kusabda) and where does the good word (susabda) reside? On what side (mukha) does the vital breath of twelve fingerbreadths reside?

M: The good word swallows or catches the bad word and itself resides within (nirantar); the vital breath of twelve fingerbreadths is controlled (rahai) through the word of the Guru.

G: Who is the Adiguru? Who is the husband of the earth (dhartri)? What is the home of knowledge (gyana)? Which is the door (duvar) of the void (sunn)?

M: The eternal beginningless (anadi) is the Adiguru; heaven (anbar) is the husband of earth; Awake-awareness (Chetan) is the home of knowledge, and realisation (parcha) is the door of the void.

G: Through the realisation (parchai) of what is the attachment with the Illusion (maya moha) broken; how can the residence of the moon (Sisihar) be pierced; how is the dam (bandha) applied and how can the body (kandha) be made immortal (ajar var)?

M: When realisation (parchai) comes to the mind, attachment to the world ceases; with the control of the working of vital breath the moon (sisihar) is destroyed; the acquisition of real knowledge (gyana) applies the dam and the realisation of the teacher (Guru parchai) gives us immortality.

G: Where do mind, vital breath, the word (shabda) and the moon reside?

M: The mind resides in the heart (hirdai); vital breath resides in the navel; the word (shabda) resides in the will (ruci); the moon resides in the heaven (gagan).

G: If there had been no heart (hirda) where would the mind have rested, composed? Had there been no navel where, would have vital breath rested unmoved? Had there been no form (rupa) where would have the word (Shabda} resided? Without a heaven where would have the moon been?

M: Without the navel, the air would have resided in the formless (Nirakar); without will, the word (shabda) would have resided (rahata) in the unmanifest (Akula); without the heaven, the moon would have resided in desire (abhika).

G: Had there been no night, where would the day have come from? Without the day, where would the night merge? When the lamp is extinguished, where does light dwell?

M: Without night, the day would have merged into Sahaj; had there been no day, the night would have passed into (Sahaj); on the extinguishing of the lamp, the light passes into the omnipresent (nirantar); had there been no vital breath, then the body of vital breath (pran purus) would have resided in the void.

G: Who is the creator of the body (kaya); wherefrom has light (tej) been created? What is the mouth (?muha or muda?) of Divine knowledge (Brahma Gyana)? How can the Unseeable be seen?

M: The Absolute (Brahma) is the creator of the body (kaya); out of truth (sat) has effulgence (tej) been created; the void is the mouth (muda or muha) of Divine knowledge (Brahma Gyana); and through the Sadguru and the disciple realisation my the unseeable be made visible.

G: How many lakhs (hundreds of thousands) of moons are there in the body?. How does fragrance reside in the flower? Where does the ghee hide in the milk? How does the soul (jiva) conceal itself in the body?

M: There are two lakhs of moons in the body; fragrance is the conscious(ness) (chetan) in the flower; the ghee is immanent in the milk; the soul (jiva) is the all-pervasiveness in the body.

G: Had there been no body where would the sun and the moon have resided? Had there been no flower, where would the fragrance have been? Had there been no milk where would the ghee have been? Had there been no body, where would the spirit have been?

M: Without the body, the sun and the moon would have been omnipresent; without the flower, fragrance would have dwelt in the (Anhad); without the milk, the ghee (ghiv or ghee) would have resided in the void; without the body, the spirit would have been in the Supreme Void (Param Sunn).

G: Where do the moon and the sun dwell, where the essence, the root of the word (nad) and the vital power (bind)? Where does the Hamsa (swan) mount up for drinking water? To what place (ghar) do you bring the reversed power (ulti shakti) to rest?

M: The moon resides in the upper (urdha) and the sun in the lower (ardha); the essence, the nad(a) and bind(u) dwells in the heart; to the heaven goes the swan (hans) for drinking water, and the reversed power (Shakti) reverts to the Self, its real, original home.

G: Where does nad(a) rise; where does it acquire equipoise (sam); how is it made to stand still, and where is it finally merged?

M: Nad(a) rises from the Unconditioned (Avigata); gains equipoise in the void; you can stop it through the vital breath and it vanishes, unites with the Formless (Niranjana).

G: If the nad(a) sounds not, if the power acts not, if the heaven is not there to draw our hope, were there neither nad(a) nor bind(u), then where would the man of vital breath (Prana Purusha) reside?

M: Nad(a) sounds, bind(u) moves; the heaven (gagan) attracts desire; but were there neither nad(a) nor bind(u) then breath would reside in the omnipresent (nirantar).

G: When form dissolves and the Formless remains, when water becomes air, when there is neither sun nor moon, where does the Hamsa dwell?

M: The Sahaja hans(a) resides after the play in the Person of the void (Sunn hans); when the form becomes Formless then the spirit (hans) resides in the Supreme light (Param Joti).

G: What is the root (mula) of the rootless (Amul)? Where does the root reside? Who is the Guru of the goal (pada)?

M: The void is the root of the rootless; the root resides in the omnipresent (Nirankar); the Guru of the goal is liberation (Nirban).

G: Where does the vital breath (prana) rise? Wherefrom does the mind come? How is the speech (vacha) born and where does it dissolve (viliyate)?

M: The birth of the mind is from the Formless (Avagat), the vital breath from the mind, and speech from the breath; speech is dissolved in the mind.

G: Which is the lake and which the lotus? How can we ward off Kal(time or death)? How can we reach the Unseeable, Unreachable (Agochar) world?

M: Mind is the lake and air is the lotus; by becoming upwards-faced (Urdh-mukhi) you can ward off Kal; through knowing the lower and the upper one my become one with the Unknowable (Agochar liv lahai).

G: Which is the difficult, and which the easy; what is union (sandh), and through what nerve centre (chakra) can the moon be made stable? How can the conscious mind attain to self-transcendence?

M: The Pure (Anila) and the Stainless (Vimal) are the difficult and easy forms of union (sandh); the dam is to be applied above the chaki nerve-centre (chakra); the always-awake can attain to self-transcendence (unmani).

G: How came about birth? How did the first consciousness begin (ad ki surat)? How was I born?

M: As oil is in the sesame seed, as fire is in the wood, as fragrance in the flower, so too resides the spirit (devata) in the body (deh).

G: What drives ahead the conch-like (sankhini)? Where does the elixir in the arched vein (banka nala) go? As the breather goes to sleep, where does the vital breath (prana) in the body (pinda) side?

M: True spontaneity (sahaj subhai) can drive the conch-like (sankhini); the arched vein (banka nala) resides in the navel; as the breather goes to sleep; the vital breath in the body resides in its own shadow or resides undivided (api chhaya or aparchhinna).

G: At what nerve centre (chakra) is the moon stabilised? At what nerve centre (chakra) is the union or penetration (sandh) applied? What nerve centre (chakra) controls (niredha) the vital air? What centre (chakra) imparts knowledge (pramodh) to the mind? At what centre (chakra) should attention (dhyana) he centred (dhariye)? At what centre (chakra) should one rest?

M: The higher (Urdh chakra), the lower (Ardh chakra), the Pashchima (west) centre, the heart centre, the throat (kantha) centre the Gyana (Agya) chakra.

G: Which is the garden, the town and the mandal? In which city is the Guru? If I forget it, how am I to cross over?

M: Whoso gives up speech has achieved the void of the manifoldness (maya sunn): in contemplating that, one rises above good and evil; by an understanding of Shiva and Shakti, one may attain to self-transcendence (unmani).

G: By what stalk of the lotus (nala) can the liquid (Shiva) be drawn up? How does the soul (jiva) drink it? How can one residing in the womb of the mother, drink the elixir?

M: It is collected through the Shankhini Nadi; the soul (jiva) resides in the Sushumna nadi; while residing in the womb of the mother he drinks it through the banka nala.

G: What is the house (graha) and what the habitation; in what womb does he remain for ten months? Through what mouth does he drink water and through what mouth, milk? In what direction was the body born?

M: The Pure, the Formless (Anil) is the house (graha) and the Unconditioned (avagat), the habitation (has); in the womb of the Beyond (Atit) he remains for ten months; through mind he has water and through vital air (pavan) he drinks elixir or milk (amrit); in the direction of Omkar the body takes birth.

G: In what void (sunn) is he born? In what void (sunn) is he absorbed?

M: He is born in the Sahaja Void and the Satguru gave him instruction at the void of nearness (Samip Sunn); he then got absorbed in the void of unattachment (Atit Sunn). He then explains to you the essence of the supreme void (Param Sunna).

G: How can one attain to Samadhi? How can one get rid of the disturbing factors (upadhi)? How can one enter the fourth (Turiya) state? How can one make one's body (kandh) changeless and deathless?

M: The young person (bala) enters Samadhi through the mind; he gets rid of the disturbances through the vital breath (pavan); he acquires the fourth state (Turiya) through attention (surat) or realisation (Gyana) and through obeying, turning to, the Guru (Guru mukh) he attains to immortality.

G: Who sleeps, who wakes, who goes to the ten directions? Wherefrom does the vital breath arise? How does it bring sound from the lips, throat and the palate (talika or taluka)?

M: The mind, or the absorbed (liv) sleeps; the vital breath or the conscious (chetan) awakes; imagination (kalpana) goes out to the ten directions. From the navel the air arises, it rises and produces sounds from the lips, throat and the palate (taluka).

G: What is the conscious? What is the essence (sar)? What is sleep and what is death? By realising what (parchai) does one sustain the elements (tat)?

M: The light (Joti) is conscious; fearlessness is the awakened essence. Waking is birth and sleeping is death; the five elements dwell in light.

G: Who speaks, who sleeps; in what form does he seek himself? In what form does he remain the same through the ages?

M: Word (shabda) speaks; attention (surat) sleeps; he seeks himself in an Unseeable (adekh) form and in the Form without Form he remains the same through ages.

G: How does the mind acquire virtues? How does the vital

air come and go? How does the fountain (nihjar) rise from the moon and how does Time or Death (kal) go to sleep?

M: In the heart (hirdai) the mind acquires the many virtues; in the navel the vital air starts its coming and going (Avagavan); contemplating itself (apmukhi) he makes the fountain play; contemplating itself Time or Death goes to sleep.

G: At which void (sunn) does light reverse; from which void does speech arise; which void is the essence of the three worlds; through which void can one cross over?

M: The void of eagerness, the void of fearlessness, the void of self-mastery, and the void of detachment. (Urga, Anbhai, Prabhu, Atit).

G: Where does hunger arise and where food? Where is sleep born and where death?

M: From desire (mansa) arose hunger and from hunger, food; from food sleep and from sleep, death.

G: At which lotus does the Hamsa (hans) inhale and exhale (sas, usas); at which lotus does Hamsa rest; at what lotus does he perform worship (puja) and at what lotus does he see the Unseeable (Alakh)?

M: The navel lotus, the heart lotus, the centre (madh) lotus, the lotus Beyond (Achint).

G: What is truth? Do tell me, please, O Guru Pandit. What is the condition or direction of the mind and the breath? How can one swim across (the ocean of Samsar) with their help?

M: Progression from mere seeing, to divine or spiritual perception (dibya drisht); from knowledge (gyana) towards realisation (vigyma); the teacher and the pupil have the same body; if realisation (parcha) comes, then there will he no straying or return.

G: Wherefrom do inhalation and exhalation arise? Where does the Param Hans reside? At what place does the mind reside constantly stable?

M: They rise from the lower (Ardh); at the higher (Urdh) the Supreme Swan (Param hans) resides; in the Sahaja Void the mind is ever in equipoise. In the realisation of the word (Shabda parchai) the mind remains in equipoise.

G: How should one come, how go; how to collect oneself and remain absorbed; how can one stabilise one's mind and one's body?

M: He should come and go in the void and in the void (sunn) he should collect himself and remain absorbed; in the Sahaja Void the body and the mind remain unchanged.

G: Where does Shiva reside and where Shakti? Where resides vital breath (prana) and where the embodied being (Jiva)? At what place can one have the realisation of them?

M: At the lower (Ardh) resides Shakti and at the higher (Urdh) Shiva; inside resides vital breath (prana) and further inside the embodied being (Jiva); by going still further in, one may attain to a realisation of them.

G: How should one sit and how walk, how speak and how meet; how should one deal with one ' s body?

M: He should sit, walk, speak and meet awake and aware (surat mukh); with his attention and discrimination (surat or nirat) thus handled, he should live fearlessly.

G: What is the word (shabda); what is attention (supat); what is discrimination (nirat)? What is the dam? How can one remain stable amidst duality?

M: The Beginningless, the Soundless (Anahad) is the word; right awareness is attention (surat); independence (niralanb) is discrimination (nirat); let him apply the check; he will then live as Unity amidst Duality.

G: Who can tread a path without feet? Who can see without eyes? Who can hear without ears? Who can speak without words?

M: Contemplation (vichar) can tread without feet; discrimination (nirat) can see without eyes; attention (supat) can hear without ears; the Soundless (Anhad) can speak without words.

G: Which posture (asan)? What knowledge (gyana)? How should the young disciple (bala) meditate (dhyan)? By what means can he enjoy the bliss of the Unconditioned Being (Avagata)?

M: Contentment (santokh) is the posture (asana); contemplation is the knowledge (gyana); he should try to rise above his physical being in (or for) his meditation; through carrying out the behest of the Guru he can have the joy of the Unconditioned Being (Avagata).

G: How to have contentment and contemplation and meditation that goes beyond the physical? How can I bend my mind to them?

M: Contentment comes from fearlessness (nirbhai); thinking from avoidance of attachment or realisation (anbhai); he should meditate within his body to rise above the body; by turning to the Teacher (Guru) one can bend one's mind to them.

G: What is the cleansing (Dhoti)? What is conduct (Achar)? Through what recitation (Japa) does the mind come to rid itself of restless thoughts (Vikaras)? How can one become unattached and fearless?

M: Meditation is purification; right thinking, discrimination leads to right action; through the Ajapa Japa (= Ha Sa) the mind rids itself of restless thoughts; by becoming unattached one can become fearless.

G: Who is the Omkar and who is the Self (ap); who is the mother and who is the father? How can the river (darya) enter the mind?

M: The word (Shabda) is the One (Omkar); light (Joti) is the Self; the void (Sunn, Surat) is the mother and consciousness (Chit or Chaitan) is the father; steadiness (nishchai—without anxiety) causes the river to flow in the mind.

G: How can one carry out true living (rahini) and how can one carry on meditation? Where is the immortal elixir? How can one drink it? How can one keep the body for ever?

M: By turning to the higher (Urdha) or the mind, you can attain right living; by turning to Shakti you can achieve right meditation; by turning to the heaven within (gagan) you can have the elixir of immortality (Amirasa) and by turning to conscious activity (chit) you can drink it. By relinquishing desire, one can gain the immortal body (bidehi rahai).

G: How should one come and go; how can one defeat death? How can one reside in light?

M: One should turn to Sahaja; one should go by turning to Shakti; by becoming wingless one can eat away death; one can always reside in light by being without breath (niswasa).

G: What is body, what is breath; what Person (purukh) should I meditate upon? At what place does mind transcend Time?

M: Air is the body and the mind is the breath (force); we should meditate on the Supreme Person (Param Purukh). In Samadhi the mind goes beyond the reach of Time.

G: Which is the key and which is the lock; who is old and who is young (bala)? Where does mind remain awakened (chetan)?

M: The wordless (nih-shabda) is the key and the word (shabda) is the lock; the unconscious one (achet) is old; the conscious one is young; mind in self-transcendence (unman) is ever aware (chetan).

G: Who is the practitioner (sadhak) and who the perfected (Siddh)? What is illusion and what is magic? How can one drive away deception from one's mind?

M: Attention is the practitioner and the word is the adept; "I am" is the illusion (Maya) and "he is" is the magic (riddh). To destroy deception or duality one should reside within.

G: Which is the mould, and which is the calx of tin? Which is the ornament and how may it be beautified? How should self-transcendence (unmani) reside changeless with that?

M: Knowledge (gyana) is the mould; vital breath is the calx (beng); light is the ornament which makes it beautiful; self-transcendence (unmani) should reside with That steadfastly, unchangingly.

G: Which is the temple and who is the god (deva); how to worship it? How should one reside there with the five unholy ones?

M: The void is the temple; mind is the god; one should serve Him by being within (nirantar); with the five one should reside in self-transcendence (Unman).

G: Which is the temple, which the door; which is the image and who is the Unfathomable (Apar)? By what method cf worship can the mind transcend itself (Unman rahai)?

M: The void is the temple; Shabda is the door; Light is the image; the Flame (Jvala) is the Unfathomable (Apar); through turning to the form of the Formless (Arupa) or to the Guru one can reside self-transcendent, or fathom the secrets.

G: Which is the lamp and which is the light? What is the wick wherein the oil resides? How can the lamp be made inextinguishable?

M: Knowledge (Gyana) is the lamp; the word (shabda) is the light (prakash); contentment (santokh—santosha) is the wick in which the oil resides; one should destroy duality and be without partiality (akhandit).

G: What goes slow and what goes fast? Who revolves and what is the find? In what place can one be fearless?

M: Steadiness of mind (dhiraj) goes slow; restless thought (vikara) goes fast; surat (attention) revolves (phurti) and truth (sar) is the find.

G: Who is a Yogi? How should he live in equipoise? Who is an enjoyer (Bhogi) and how should he acquire? How does pain rise out of pleasure and how can one patiently suffer pain?

M: Mind is the Jogi; let him live in self-transcendence; the great elixir will come to him and he will enjoy all pleasures; in that elixir is the indivisible (akhandit) pain; the word (shabda) of the Guru secures the patience to suffer it.

G: Which is the self (Atma), what comes and goes? Which is the self, what is absorbed in the void? What is the self, what stays changeless in the three worlds? By knowing whom can one become one of the fifty-two heroes (bavan bir)?

M: The self of vital breath (pavan atma) comes and goes; the mind-self is absorbed in the void; the knowledge-self resides unchanged in the three worlds; by knowing (parchai) the Guru one becomes one of the fifty-two heroes.

G: What is the life (Jiva) of the mind? What is the support (besas) of that life (jiva)? What is the basis of that support? What is the form of the Brahma?

M: The life (jiva) of the mind is the vital air and the support of the embodied being (jiva) is the void; the basis (adhar) of that support (besas) is the form of Brahma (= Absolute); and the form of the Brahma is unthinkable (Achintya).

G: Through which centre can one make one's body immortal? Through which centre can one attain to the Unknowable dam (Agochar bandh)? Through which centre can the Hamsa be liberated (Hans nirodh)? Through which centre can the mind be instructed? Which centre gives pleasure? Which centre brings on the Samadhi?

M: The anus (Mula) centre; the penis (Guda) centre; the navel (Nabhi) centre; the heart centre; the throat centre and the head (Nilata) centre. He who knows the meaning of these six chakras, he is the Cause, he the Deva.

He is a Yogi who controls (sadhita) the mind and the vital air (pavana); he is not stained by evil. He is not seized of merit.

The Jnana Karika

The Jnana Karika, like the Akulavira and the Kulananda Tantra, is another text ascribed to Gorakhnath's guru Matsyendranath, published in the original version of Bagchi's Kaulajnananirnaya of the School of Matsyendranath. The verse numbering in the original Bagchi version was out of kilter and this has been corrected. The text is presented in itrans format below.

Jnana means knowledge, while Karika means a set of verses explaining the philosophical doctrines.

Divided into three brief chapters dealing with liberation, dharma and adharma, and the way of acting or the path, it is of considerable interest to the whole Nath tradition. The third chapter gives the inner meaning of a host of terms associated with Tantra—some of which continue to mislead because the often bizarre and uncanny terms are taken at face value.

In chapter one, which has 18 verses and some lacunae, liberation, that is knowledge or jnana, is described in terms more of what it is not, than what it is. For example, it is devoid of the 25, it is all peaceful, free of all attachment, free of mantra and Tantra or distinctions of other types, free of the dualities, without Kulas (Shaktis). People wander in samsara deluded by the stain of Maya, says Matsyendranath. But mental attachments, devas and devis, mental attachments and so forth are nothing to do with liberation. The state of liberation is devoid of divisions, attachments, tattvas, different spheres of activity and any other mental state.

Chapter two talks of this Jnana further, which, the text starts by saying, is free of the Sun and the Moon, and the five elements. It pervades all these, just as ghee is present in latent form in milk. It is also free of the three gunas, of manas and the other mental conditions, and is devoid of both thinking and non-thinking, the highest of the high, and is free from the elements, the sense impressions (indriyas), buddhi, ahamkara or manas. Compared to this motionless, timeless state, concepts such as dharana, dhyana, the sun, moon and fire, as well as the elements, Brahma and the other devas resemble false imaginings.

The tantrik is often enjoined to do his or her puja at a place where is a single Shiva lingam, a cremation ground, the junction of two rivers, a barren place, at the root of a tree, at a crossroads, next to a great ocean, or at the junction of three paths. She or he is to be naked, with dishevelled hair, intoxicated by liquor.

Chapter three Jnana Karika gives an internal explanation of these symbols. The lingam, according to this text, is an internal one, and not one made from stone, silver or gold.

Siddha Matsyendranath is further made to say that the cremation ground is the union of inspiration and expiration. The prana or vital breath moves through the body, and is the basis of the bioenergetic web. The movement of prana above is the Ganges, while that below is the Yamuna. The junction of the two rivers is the union of inspiration and expiration.

The text, referring to desolate places for sadhana, says that the true essence of prana is voidness. The tree is the human body, the root of the tree is therefore the central point of the upper and lower streams, each of which branches out into other nadis or conduits of bioenergy. The crossroads means the central point of orientation, beyond the four elements, represented by the Shaktis Jaya, Vijaya, Ajita and Aparajita. These four elements are within the body.

When a sadhaka is told to go to the edge of a great ocean, this text says the meaning of this is the bindu, or point into which all the lesser nadis flow. This ocean has a coast or boundary which separates Sun from Moon.

The three paths are the three gunas of Sattvas, Rajas and Tamas. The sadhaka is to be naked, this being the sumbolism of a simple and free soul. Dishevelled hair shows freedom from dharma and adharma. Intoxication with wine means being full of the bliss of the nectar of yoga.

Shri Nathanavaratnamalika

This tiny Sanskrit work, the rosary or garland of the nine gems of the Nathas, is found in the 1953 Ganesha & Co version of Sir John Woodroffe's translation and text of the work on Shri Vidya Kamakalavilasa. Ascribed to Maheshanatha, the text includes a commentary by the renowned Shri Vidya upasaka Bhaskararaya (Bhasuranandanath).

The briefness of the text, reproduced in itrans format below but without Bhaskaraya's commentary, belies its importance, as it deals with the number symbolism of nine and how this relates to the 21,600 breaths a human is supposed to take in a day, as well as the identity of these with the matrikas, or letters of the Sanskrit alphabet, the Shri Yantra, and time itself.

As the first shloka says, Hamsah is the gayatri mantra produced by the breathing, and this is one with the unconscious recitation of the mantra so.aham, pervading all human beings.

Sir John Woodroffe says in a commentary to his translation of the Anandalahari (Wave of Bliss), published in 1916: "Shiva can do nothing without Shakti which is of threefold aspect of Iccha (will), Jnana (knowledge) and Kriya (action). The author here speaks of the Mantra Hamsah. Ham is the Bija of Shiva and Sah that of Shakti. Ham+Sah = Hamsah = Sah+Ham = So Ham = So'Ham = Sa+Aham, So Ham being Sah+Ham = Shakti+Shiva; if S and H be eliminated therefrom there remains Ong or Om the Pranava..."

The Navanaths of the title of this work are, in the Tantrarajatantra, linked to the nine orifices of the human body, and to the nine mandalas of the Shri Yantra.

As a human being, in these schools, is considered as a microcosm, the in-breathing and the out-breathing symbolise the creation and the dissolution of the universe. The realisation of Ha+Sa, Sun and Moon, Shiva and Shakti, in-breathing and out-breathing is to become one with the universe itself.

But this, according to these schools, cannot be achieved without a yogic understanding of the other effects of the wheel of time, one, as the Yoginihridaya states, with the mandalas of the Shri Chakra, the letters of the alphabet (sound/mantra), and the Shaktis or attendants of the goddess.

Some of these Shaktis, as the introduction to the Malinivijayottaratantra have the function of preventing such a realisation, while others foster this. Further, according to various texts and commentaries of Kashmir Shaivism, ignorance and other defects also prevent the realisation of one's essential unity with Shiva-Shakti. The normal course of creation is pravritti, an expansion or flowing outward. The sadhaka is to cultivate nivritti, ulta sadhana—a reverse movement, or kaya sadhana—cultivation of the body.

This may have little or much to do with ritual worship (puja), which if performed without an inner realisation of the principles it embodies is considered to be mummery.

The different nyasas of the Shri Vidya tradition are intended to bring to a sadhaka the realisation of his or her essential unity with the matrikas, with the constellations (rashi) planets (graha) which includes the Sun and the Moon) and the 27 asterisms (nakshatra), and breath itself.

Pratised ways to achieve this realisation are reputed to be the inner tradition of sadhana taught in some schools, and which may include a number of different methods. Some of these may require an intense struggle because an individual, not realising that she or he is Shiva-Shakti, instead identifies with partial aspects or Shaktis.

The Tantrarajatantra hints at some of these methods, such as the way the grahas or planets influence the breath and therefore prevent this realisation as they affect the musculature and other parts of the human bionergetic web. Freeing oneself from these misidentifications also frees up the natural flow of Pranashakti (herself one with the supreme Devi), in the body.

A teacher who understands the movement of these currents (nadis) and the relationship between the wheel of time (Kalachakra), divinity, and the body itself is said to be a requisite in these different tantrik schools.

Siddha Siddhanta Paddhati

This Sanskrit text, attributed to Siddha Gorakhnath, is divided into six chapters called Upadeshas. These sections are 1) origin of Pinda, 2) discussion of Pinda, 3) knowledge relating to Pinda, 4) foundation of Pinda. 5) unity of Pinda with the Supreme Reality (Parampada), and 6) the nature of the Avadhoot.

The Parampada is also known as Anama, or the nameless. The Pinda itself is Shakti. Pinda means, literally, a ball or an egg. This egg is the cosmic egg or Macrocosm and also the microcosmic egg, or the human being. It has six forms, called in this text Para (Supreme). Anadi (Without Origin), Adi (Origin), Mahasakara (Great Body), Prakrita (Natural Body) and Garbha (Womb-born Body).

Each of these six aspects of Pinda has itself five factors, these being subdivided into five other divisions. So each of the six aspects of Pinda has 25 qualities.

The five divisions partake of the nature of Space, Air, Water, Fire and Earth—the five elements or Bhutas. This work primarily belongs to the Kanphat or Gorakhnathi tradition, and having many contacts with the Adi Nath tradition, should be compared with Kaula Jnana Nirnayam (Prachya Prakashan, 1986). Descriptions of the chakras should not be taken at face value. Many different chakra systems exist. It is only in recent history that the one described in Avalon's Serpent Power has come into general vague.

Chapter One

The first of the six Pindas is Para, or the Supreme. This is identified with Shakti, whose 25 divisions are shown as follows:

Para Pinda: 1) Nija or indwelling Shakti, with the five qualities Eternity, Stainlessness, No Sound, No Light, No Emanation. 2) Para Shakti with the five qualities of Non-dependency, Immeasurability, No Divisions, Endlessness, Unmanifastness. 3) Apara or Manifestation Shakti, with the five qualities of Quivering, Emanation, Abundance, Distinction, Vibration. 4) Sukshma or Subtle Shakti, with the five qualities of Wholeness, All Extensiveness, Immovability, Firmness, and Changelessness. 5) Kundalini Shakti with Her five qualities of Fullness, Reflectiveness, Mightiness, Power and Openness.

Anadi Pinda: 1) Parampara or Uninterrupted Line, with five qualities of Spotlessness, Without comparison, Beyond all, Without form, Never appearing. 2) Param Padam or Supreme Part with five qualities of No Parts, Very Highest, Without Movement, Numberless, Supreme. 3) Shunya or Void with the five qualities of Playfulness, Fullness, Agitatedness, Unsteadiness. Fickleness 4) Niranjana or the Stainless, with the five qualities of Truthfulness, Spontaneity (Sahaja), Perfect Assimilation (Samarasa), Attentiveness and Omnipresence 5) Paramatma or Supreme Being with five qualities of Imperishability, inability to be Divided, Inability to be Cut, inability to be Burnt, inability to be Destroyed.

Adi Pinda: 1) Paramananda or Supreme Bliss with five qualities of Vibration, Happiness, Power, Quietude, Eternal

Bliss. 2) Prabodha or Manifestation with five qualities of Arising, Growth, Shining Forth, Expansion, Light. 3) Chidudaya or Arising of Consciousness with five qualities of Good Meditation, Discrimination, Doing, Knowing, independence. 4) Prakasha or Illumination with the five qualities of being Undisturbed by Things, Completeness, being Unaffected by Thought, Sama or Equipoise, and Relaxedness. 5) So-Aham or That I Am with five qualities of lmmortality, Entireness, resting in one's own Atma, Cosmic Meditation and Equality with All.

Mahasakara Pinda: 1) Maha Akasha or Great Space, with the five qualities Space, Intactness, Untouchability, Consisting of the colour blue, relating to Sound. 2) Maha Vayu or Great Air, the five qualities being Moving About Trembling, Touch, Drying, consisting of the colour purple. 3) Maha Tejas or Great Fire relating to Burning, Cooking, Heat, Sight, and the colour red. 4) Maha Salila or Great Water with the five qualities of Flowing, Moistness, Liquidity, Taste, and the colour white. 5) Maha Prithivi or Great Earth, with five qualities of Grossness, Different Bodies, Firmness, Smell, yellow.

Prakrita Pinda: 1) Earth with the five qualities of Bone, Flesh, Skin, Veins and Hair. 2) Water with the five qualities of Saliva, Sweat, Semen, Blood, Urine. 3) Fire with the five qualities Hunger, Thirst, Dream, Languor, Idleness. 4) Air with the five qualities Running, Swimming, Stretching, Bending, Disappointment. 5) Earth with the five qualities of Disease, Hatred, Fear, Shame; Delusion.

The work then proceeds to give five-fold qualities of many other things which seem to pertain to the Garbha Pinda. They are enumerated below.

Qualities of the Antakarana: The Antakarana is the inner complex carried from birth to rebirth. 1) The five qualities of Manas (mind) are Resolution, Wavering, Folly, Stupidity, Mentality. 2) The five qualities of Buddhi (reason) are Discrimination, Dispassion, Peace, Contentment and Patience. 3) The five qualities of Ahankara or Ego are Wishing to have contact, the feeling "this is mine", My Happiness, My Sorrow, This is Mine. 4) The five qualities of Chitta or Observation are

Pondering, Constancy, Memory, Reflection, and Making one's own. 5) The five qualities of Chaitanya or full awareness are Reflectiveness, Skill, Steadiness, Thoughtfulness, and Indifference.

The Five Kulas: 1) Sattvas—the five being Compassion, Duty, Mercy, Devotion and Faith. 2) Rajas the five qualities Giving, Enjoyment, Eroticism, Possession, and Having Wealth. 3) Tamas with the five qualities Argumentativeness, Grief, Quarrelsomeness, Bondage and Fraud. 4) Kala or Time with the five qualities Divisions, Periods, Movement, Measure, and Lack of Substance. 5) Jiva or Embodied Being with the five qualities Wake, Dream, Deep Sleep, the Fourth, and that beyond the Fourth.

The Five Shaktis of Manifestation: 1) Iccha, with her five qualities Divine Madness, Desire, Longing, Reflection, and Achieving what is Desired. 2) Kriya, with the five Making Love, Effort, Action, Steadiness, and adherence to one's own Kula-cluster. 3) Maya with her five qualities of Arrogance, Envy, Deceit, Acting, and Playfulness. 4) Prakriti with her five qualities Hope, Thirst, Eagerness, Wishing, Duplicity. 5) Vak or Devi as Speech with the five qualities Supremacy, Pashyanti, Madhyama, Vaikhari and Matrika.

The Five Gunas of Personal Experience: 1) Karma, the five being Good, Evil, Fame, Dishonour, Looking to the results of action. 2) Kaama or sexuality with the five qualities of intercourse, Liking, Playfulness, Desire or Lust. 3) Moon with 16 Kalas, and a 17th called Nirvana. 4) Sun, with 12 Kalas and a 13th called Shining by its own Light. 5) Fire with 10 Kalas, the 11th being Light. (In these last three gunas the 17th, 13th and 11th Kalas are each taken as a synthesis of 16, 12 and 10).

Channels of Bioenergy (Nadis): These are enumerated in the text as Ida, Pingala—both of which are related to the nostrils; Sushumna, which is the central channel, Saraswati, which is on the tongue; Pusha and Alambusha related to the eyes; Gandhari relating to the hands and the ears; Kuhu, which goes to the anus; Shankhini, said to be the lingam aperture. The Brahmarandhra is related through the central path to all of the 10 Nadis.

The 10 Vital Breaths or Vayus: These are related to different functions in the body. The most important vital breath is Prana, said to reside in the heart and consisting of expiration and inspiration, relating to the letters Ha and Sa (Hamsa).

The rest of the first Upadesha describes how, by the combination of red blood and white semen, birth occurs, and enumerates the different stages in the development of an embryo. It is stated that an excess of semen gives males, blood females, and an equal amount gives rise to neuter, hermaphrodite, or homosexual. The chapter closes with the proportions of the different Ayurvedic bases in the body, and states that Vata, Pitta and Sleshma—the three base Dhatus, give rise to the 10.

Chapter Two

This section deals with the position of the chakras in the body. The fundamental chakra is the place of Kamarupa, it is of a wine-colour, giving the fruit of all sexuality. Shakti is said to reside here. The second chakra is called the Svadishtana, in its centre is a lingam the colour of pink coral, like a young shoot. In there is Oddiyana Pitha, giving the power of all attraction.

Thirdly is the navel chakra, with five petals, and in its centre is Kundalini Shakti coiled up. She is said to resemble 10 million dawn suns, and gives all siddhi. The fourth chakra is the heart-centre, with eight petals. In it is a lingam. It is the seat of Hamsa, the place where all the senses come to reside.

The fifth is the throat chakra, the junction point of Ida and Pingala. Ida is the Moon nadi on the right, and Pingala the Sun nadi on the left. In the centre is Sushumna. One should meditate there on spontaneous sound, which is Nada. Above this is the Talu chakra. Amrita is said to flow from here. It is near the uvula. It is called Rajadanta, and is said to be the place Shankhini Nadi comes to the 10th door or aperture. One is to meditate there on the Void.

Above this is the brow chakra, said to be the Eye of Knowledge. One obtains Siddhi of the circle of the Matrikas by meditating here. It is like the source of Light. The eighth chakra is said to be the Brahmarandhra or Nirvana Chakra. It is the colour of a column of smoke (purple). The three Kutas or peaks are above this. Jalandhara is situated there. If one meditates on this centre, it gives liberation. Above this is another chakra called the Akasha or Space Chakra. It has 16 petals, and in its centre is an Upper Yoni. Over this one should meditate on the Supreme Void, which is said to be the place of Purnagiri Pitha. It gives all desired siddhi.

The 16 Adharas: The text now mentions 16 places where meditation may be accomplished. On the tip of the big toe of the right foot one should meditate on a steady light. The second base is situated in the root chakra, and a flaming fire should he visualised there. Thirdly is the anus, where the Apana vital breath resides. The Fourth is in the penis, where the Brahmagranthis are said to come together. Fifth is the Oddiyana base. Sixthly is the navel centre, in which is Om, where all sound dissolves. The seventh is the heart chakra, where Prana resides.

The eighth is the throat adhara, the place where Ida and Pingala come together. The ninth base is the Ghantika, at the root of the tongue, whence arises the nectar. The 10th is behind this, identified with the Talu chakra. The 11th base is at the tip of the tongue. Meditating here one conquers all disease. The 12th centre is the third eye, where one should meditate on the lunar circle.

Next and 13th is the spot at the root of the nose. Meditating here, one becomes very concentrated of mind. The 14th base is behind the root of the nose. The 15th is on the forehead, and is said to be the centre of Light. At the 16th, above the Brahmarandhra, is the Space Chakra, and here reside the two lotus feet of Shri Guru.

Three Lakshyas or Places of Meditation: These are identified with Moon, Sun and Fire at head, heart and genitals.

Five Spaces: These pervade the body, and each has the characteristic of Voidness. It is stated here that only by meditating on the nine chakras, the 16 bases, the three Lakshyas and the five Spaces does one become a yogi. In passing, it should be noted that the Kashmir Shaivite Netra Tantra follows the above scheme most closely. The chapter closes with a description of the well known eight limbs of yoga.

Chapter Three

This section deals with the identity of macrocosm and microcosm. The tortoise supporting the cosmos is below the feet, On the soles is the Patala underworld. Talatala is in the region of the front of the feet, Mahatala is on the heels. Rasatala is at the ankles. Sutala is associated with the legs. Vitala is in the region of the knees, and Atala is at the root of the body. Above this resides the Great Fire at the End of Time (Shiva Kalagnirudra).

The three worlds are then described. Bhur is in the genitals. The presiding deity is Indra. At the tip of the penis and at the penis aperture is Mahar Loka. Svar Loka is associated with the womb. In the heart is Rudra Loka. Rudra is said to be one with Ugra. The chest region is Ishvara Loka. The throat region is Sadashiva Loka. In the centre of the throat, in the neck, is Shri Kantha Loka. At the tongue root is Bhairava Loka—the Heaven of Bhairava. In the 10th aperture is Shiva Loka Above this 10th aperture is Siddha Loka, where dwell eternally the Siddha Nathas.

In the forehead is the Heaven without Origin. The Lord there is Anadi, or the Originless One. At the peak of the head is Kula Loka, the Lord there being Kuleshvara. In the Brahmarandhra is the Lord of the Supreme Absolute. In the Trikuta is Shakti Loka, and Supreme Shakti rules here.

The seven underworlds and the heavens all reside in the human body. In the nine apertures are the nine divisions (Khandhas) of India. The seven oceans are Identified with the seven bodily Dhatus. The spine is Mount Meru, and Mount Kailasha is the aperture at the top of the head. Other mountain ranges exist where there are bumps on the body. The Vindhya range is on the right ear, and on the left Mount Mainaka.

Shri Parvati is on the forehead. The 64 Yoginis dwell in the joints of the hands and fingers along with the smaller mountain ranges. The great rivers Ganges, Jumna, Chandrabhaga, Saraswati, Narmada & Caveri are identified with the veins. Other lesser rivers and streams are associated with the veins and subtle channels of energy throughout the body. Also in the body are the 27 sidereal constellations, the 12 sidereal constellations, all pithas, and the lunar days.

Dwelling in the pores and hairs of the body are the 33 millions of gods and goddesses. Numberless saints are associated with the armpit hair. The Pithas and lesser Pithas (Upapitha) reside in the facial hair. Associated with all the Joints of the body and the other places mentioned are the Elements, the Ghosts (Pretas), the Pishachas, the Rakshasas, the Daityas and the Danavas.

The Gandharvas, Kinnaras, Ganas, Apsaras and Yakshas also dwell in the body. Speech is equivalent to the rays of light outspreading in the cosmos. The Khechari Shaktis, and Dakini &c. dwell in the body. Wind is equivalent to breath, and if tears fall it is equivalent to the rain.

All the sacred bathing places are in the (108) marmas of the body. The lights of consciousness are the Siddha Naths. The Sun and the Moon are the two eyes. The sentiments reside in the hairs of the legs. Insects and other creeping things are in the urine and waste products.

When a person is happy, she or he is in heaven. When sad, it is hell. Free from these distinctions, one is liberated whether asleep or awake. Maheshvar (Shiva) dwells completely without distinctions in this Cosms, emanating it and shining forth by His own light.

Chapter Four

This section deals with Shakti, who is the support or basis for the Pinda previously mentioned. Kula is manifested Shakti, whilst Akula is non-dual, without any distinctions whatsoever. The union of Kula and Akula is called Samarasa or perfect assimilation.

Parampada may be likened to Supreme Shiva, whilst Kula is Shiva in His immanent form. Both Kula and Akula are inseparate.

Various extracts are given from tantras. These are 'Lalita Svacchanda', 'Pratyabhijna' and 'Vamakeshvara'.. The last extract is to the effect that Shiva and Shakti are one.

Other extracts from other tantras are quoted to further explain the theory behind the practice and to explain what has previously been mentioned.

Chapter Five

Mainly deals with the supremacy of the Guru, and the attainment of the equilibrium of the Pinda, which results in the achievement of Samarasa or perfect assimilation. Only through the grace of the Guru may this be achieved and not through thought or endless discussions. One should obtain it orally and not from a multitude of texts. Only then is one liberated. Parampada is obtained only through the favour of the true Guru.

One who has achieved this Samarasa alone is a Svecha Yogi, able to do whatever is willed, free from sickness and death. The results of practice for a period of years are described in the ninth year one achieves a body which is like Vajra. In the 12th year one becomes equal to Shiva, is worshipped in the three worlds, and a Siddha like Shri Bhairava. Success is not achieved by recitation of mantra, penances, meditations, sacrifices, pilgrimages, or worship of Devas, but only through the Guru's grace.

A couplet is given, said to have been spoken by Shiva: "There is nothing greater than Guru. There is nothing greater than Guru. There is nothing greater than Guru. The Guru is Shiva. The Guru is Shiva. The Guru is Shiva. The Guru is Shiva."

If one is not instructed by the Guru but attempts the great work alone then one is a liar as all is achieved through his grace. Such a person is empty of all knowledge.

Chapter Six

Deals with the characteristics of an Avadhoot—one who has achieved the highest state of all. Such a person is a Siddha Yogi, free from everything, with a complete understanding of the Pinda. Only an Avadhoot may initiate a disciple into the path of Nath Yoga. The Natha school is the best of all other systems, and therefore the Avadhoot is the best of all Gurus. Systems and paths mentioned include Sankhya, Vaishnava, Vedik, Saura, Buddhist, Jaina, and many others.

This path is so superior that it should be carefully hidden. The lotus feet of the Guru should be sought if one wishes to achieve success, and to be free from fear and sorrow.

2

Schools of Tantras

Nityotsava

The Nityotsava contains very little philosophical material on the cults which centre around the famous Shri Yantra, but a wealth of detail on the ritual, the mantras and other features of this complex tantrik form. These include not only the worship of Mahatripurasundari, but also of Ganapati (Ganesha), Dandini (Varahi) and Shyama (Kurukulla). There are also chapters dealing with practices common to all tantrik devata.

Mantras and other ritual details below use the itrans format, which you can view in their Devanagari characters by copying and pasting into the freeware programme Itranslator.

Chapter One

This section is devoted to initiation (diksha), and starts with a salutation to the nine Nathas, to Shiva, to Ganesha, to the Empress (Maharajni = Tripura) and to the gurus of the tradition.

After this, Umanandanatha goes on to describe initiation, starting first with descriptions of the act contained in the Kalanirnaya (Kaulanirnaya?) and the Manthana Bhairava tantras. This section outlines the times for diksha, which are closely linked to Indian sidereal astrology. Initiation in certain of the 27 asterisms brings different results, while the days of the week also have their own merits, while the 15 days of the moon (tithis) also are said to have specific results. The text then quotes from other texts along similar lines.

The text quotes authorities for times which are not suitable for initiation, which mentions a number of yogas (planetary combinations in this context), where malefics, that is the Sun (Ravi), Mars (Bhauma), Saturn (Shani), the nodes of the moon (Rahu and Ketu) and the Waning Moon are conjoined in certain places in a horoscope for the time of initiation.

The next section in this chapter deals with the characteristics of guru and pupil, quoting from the Tantrarajatantra. The guru should be handsome, full of virtues, situated in himself, and know the essence of many tantras. He should be free from doubt, having cut through such doubts by the grace of his own guru. The pupil should be free from greed, controlled of senses, steady, faithful, and devoted to guru, mantra and devata.

The text mentions the nature of Tripura Siddhanta, based as it is on the 35 (36) tattvas of earth, water, fire, air, space (aether), smell, taste, sight, hearing, sound, etc. Then follows details of mantra upasana, the duty of an upasaka of Tripurasundari, and different types of initiation such as Shambhavi diksha, Shakti diksha and Mantri diksha. The text talks of Samayachara, of the Kuladharma, and the competence of people for given mantras.

Chapter Two

This chapter deals with Ganapati (Ganesha) and his ritual worship, including the mantras, tarpana (oblations) and nyasas necessary.

It then goes on to deal with the puja proper, the necessary substances with which to accomplish this, the dhyana of Vighneshvara (Lord of Obstacles), worship of the pitha Shaktis, named in this text as Tivra, Jalini, Nanda, Bhogada, Kamarupini, Ugra, Tejovati, Satya and Vighnashini. Mantras should commence shrim hrim kiim and end with namah. The six limb puja (shadangapuja) is as follows:

shrim hrim kiim OM gam hridayaya namah hridaya-shaktishripadukam pujayami.

3 shrim gim shirase svaha shirashshaktishripaduka. pujayami.

3 hrim gum shikhayai vashath shikhashaktishri-padukam pujayami.

3 kiim gaim kavachaya humh kavachashaktishri-padukam pujayami.

3 glaum gaum Netratrayaya vaushath netratraya-shaktishripadukam pujayami.

3 gam gah Astraya phath astrashaktishripadukam pujayami.

The text then goes on to detail the three aughas or lines of guru related to Mahaganapati, which are divided into the divya (celestial), the siddha and the manava (mortal) lines.

The text then describes puja of the Mahaganapati yantra, which is divided into five parts. The first set of attendants relate to the three lines, the six lines, and the space between the lines. The second avarana relates to the six konas. The third avarana is connected with the junction points (sandhis) of the six konas, and are the six limb devatas.

The fourth avarana is related to the eight petals, starting from the west, with the Devis there being Brahmi, Maheshvari, Kaumari, Vaishnavi, Varahi, Mahendri, Chamunda and Mahalakshmi. The fifth avarana relates to the earth-square or bhupura and worships the guardians of the directions, the kshetrapalas, given in the text as Indra, Agni, Yama, Nairriti, Varuna, Vayu, Soma, and Ishana.

Each of these forms holds his appropriate weapon, rides his appropriate vahana, and holds his appropriate direction and sub direction in the order given. Thus, although the Nityotsava contains no illustration of the yantra, it can be constructed as triangle, hexagon, eight petals and earthsquare.

After worship of the avarana deities, Gananatha (Ganapati, Ganesha) should be worshipped with sixteen ritual accessories. Then follows a description of the fire sacrifice, the giving of bali, a stotra about tarpana (oblation), a dhyana or meditation on Ganapati, the ashtakam or eight limbed hymn, Suvasini puja, and the purashcharana, or preparatory rites to be performed in his worship.

Chapter Three

This large chapter, called Shri Krama, deals with the worship of Lalita Tripurasundari, and opens with Umanandanatha bowing to his guru, Shri Bhasurananda Natha.

After outlining the contents of this important chapter, Umanandanatha starts by describing the guru meditation, and describes a process known as pranasamyaman, where the vital energy is directed to the top of the head, where the guru is said to reside.

The sadhaka is then recommended to meditate on the ajapa gayatri, the "non-recited" gayatri of 21,600 breaths a human takes everyday. Then follow details relating to bathing, and to sandhya or the twilight worship. Here, there is to meditation on the sun as Martandabhairava, seated in union with his Shakti, Prakashashakti. The tantrik Gayatri of Tripurasundari is described as:-

aim hrIm shrim ka e I la hrIm tripurasundari vidmahe aim hrIm shrim ha sa ka ha la hrIm pithakamini dhimahi aim hrIm shrim sakalahrim Tannah klinne prachodayath.

This combines the Gayatri with the 15-lettered vidya of Lalita Tripurasundari.

After these preliminaries, the Shri Krama chapter begins to describe the puja proper, together with the rites necessary when entering a temple. These include worship of the different kalas (parts) of the Sun, Moon and Fire, followed by a short section which describes the material of which a Shri Yantra may be fashioned. Gold, silver, copper and other metallic plates are mentioned, while substances such as sindura, kumkum, and milk are required. The form of the yantra is described, but only as bindu, triangle, eight triangles, two sets of 10 triangles, fourteen triangles, eight petals with filaments, 16 petals with filaments, three circles and four lines. Next comes the procedure of breathing life into a yantra, followed by temple worship, a section which we have translated elsewhere on this chapter.

The process of bhutashuddhi, or purification of the elements follows, followed by a technique using mantras which is supposed to fence off the area from obstructions, bhutas, and other obstacles.

Nine nyasas are recommended for the worship of Lalita, and these are listed in the text as matrikanyasa, karashuddhinyasa (hand purification), atmarakshanyasa (protection of the atma), chaturasananyasa, Bala six-limbed nyasa, Vashini and the other (Vakdevatas) nyasa, root vidya nyasa, shodhanyasa and chakranyasa. The next section deals with placing of the vesel for the ordinary offering, which are to be upon a design with a bindu, a triangle, a hexagon and a square enclosure. This is followed by a long section describing the preparation of special offering.

Umanandanatha quotes from the Jnanarnava Tantra next to describe the inner yaga or meditation on the Shri Yantra which is to be performed, while the next section deals with the worship of the 64 ritual accessories (upacharas) used in the puja. Then follow the mantras of the Fifteen Nityas, which, unlike in the Tantraraja, are given in full and without special code.

A large section on the different gurus in the Shri Vidya tradition follows this section, and, as in the Ganapati section of Nityotsava, these are divided into celestial, siddha and mortal gurus, both for the vidya (mantra) that begins with Ka (Kadi) and for that which begins with Ha (Hadi).

The celestial gurus in the Kadi group are listed as Paraprakashandanatha, Parashivanandanatha, Parashaktyamba, Kauleshvaranandanatha, Shukladevyamba, Kuleshvaranandanatha and Kameshvaryamba. The siddha aughas are Bhoganandanatha, Chinnanandanatha, Samayanandanatha and Sahajanandanatha. The manava augha consists of Gaganandanatha, Vishvanandanatha, Vimalanandanatha, Madanandanatha, Bhuvananandanatha, Lilanandanatha, Svatmanandanatha and Priyanandanatha. At the close of this lengthy section, mantras are given for other, unknown gurus.

The next section opens with the puja of the attendants of the mandalas of the Shri Yantra, nine in number. As these have been listed in full elsewhere on this chapter, we refer you to the appropriate sections. This is followed by the sadhaka being enjoined to meditate on Kamakala, which consists of the three bindus and the ardha-matra. Brief instructions for homa, giving bali, pradakshina and circumambulation follow, before a stotra or hymn to Tripurasundari is to be recited.

Following this is Suvasini Puja, with a mantra given for purification of the Shakti which reads aim hrim shrim aim kiim sauh tripurayai namah imam shaktim pavitri kuru mama shaktim kuru svaha. The Shakti should then be given various good things such as garments, flower, incense, light, unguents and powders, as well as food and pan. A lengthy section dealing with purification of wine and other substances used in the rite should follow.

Although mentioned briefly above, a lengthy section follows dealing with the right way to create a Shri Yantra, and the number of marmas and sandhis that result from drawing it correctly. This is so interesting that in future we will place it on the chapter in itrans format and translate it. Then follows a brief section of the different prastaras of the Shri Yantra and the pranapratishta of the Shri Chakra, which imbues the device with life and which is translated elsewhere on this chapter. A yantra made of gold lasts for life, one of silver 22 years, copper 12, and on bhurja bark six years.

A lengthy section follows on the homas that may be performed, while the next section describes the mudras or hand-gestures used in the Kadi Shri Vidya cult.

The next section deals with the types of nyasas used when worshipping Tripurasundari. These were mentioned earlier and a number of them, including the lengthy Shodhanyasa and the Shri Chakra Nyasa are elsewhere on this chapter.

Instructions on how to perform japa (recitation of the vidya) are then given. Instructions are given for reciting various vidyas such as the Kameshvari mantra, the Kamakala mantra, the Utkilana mantra, and a large number of other mantras relating to Tripurasundari, to her aspects or to various elements of her worship, such as the rosary, are then described in full.

The Fifteen Nitya mantras are given again, as well as special mantras for "miraculous" aspects of Tripura such as Asvarudha, Bala, Annapurna, Svapnavarahi and many others. Some of these aspects, together with the prayogas (applications) relating to them are described in the Tantrarajatantra and other places, and relate to magical powers that a sadhaka can achieve through their use. A section then follows on optional rites that may be performed in various solar months, and which the Tantrarajatantra also details.

Chapter Four

Details the rites of Shyama, the dusky form of Tripura, who is described elsewhere in Shri Vidya tantras as Tara or Kurukulla, and who is the "mother" form, just as Varahi is the "father" form of the goddess. The nyasas and the other elements in tantrik puja are described, together with the yantra to be drawn, and the avaranas to be worshipped.

In the central triangle Rati, Priti and Manobhava receive puja, followed by the five arrows in the triangles of the pentangle. On the tips of this star the five forms of Krishna, one with Kali, receive worship, while the eight devatas Brahmi, etc. receive worship in the eight petalled lotus. On the tips or filaments of the lotus Lakshmi, Saraswati, Rati, Priti, Kirti, Shanti, Pushti, and Tushti are worshipped. The sixteen petalled lotus has sixteen Shaktis or attendants named in the text as Vama, Jyeshtha, Raudri, Shanti, Shraddha, Saraswati, Kriyashakti, Lakshmi, Srishti, Mohini, Pramathini, Ashvasini, Vichi, Vidyunmalini, Surananda and Nagabuddhika. This is the fourth avarana.

In the outer lotus of eight petals are the Bhairavas Asitanga, Ruru, Chanda, Krodha, Unmatta, Kapali, Bhishana and Samhara. Inside the four petals are Matangishvari, Siddhalakshmi, Mahamatangi, and Mahasiddhalakshmi. This is the sixth avarana. Ganapati, Durga, Vatuka and Kshetrapala are to be worshipped in the square, followed by protectors of the directions.

Then follows a description of recitation of the mantra of Matangi, followed by praise of that aspect of the goddess. Rules for an upasaka (worshipper) of Shyama follow, including details of the purashcharana and so forth required, as well as the types of homa, a description of the Kurma Chakra, purification of the mala or rosary, and a lengthy section on how to purify Rudraksha malas specifically. Optional homas and the dimensions of the fire pits to be used are detailed.

Chapter Five

This chapter of Nityotsava deals with the puja rites of Dandini, in this context similar to the "father form" Varahi. As in previous chapters, the various nyasas, upacharas and other ritual details relating to Dandini are described, together with the gurus of the parampara and the different avaranas of her yantra. The mantras of Varahi and a lengthy Varahi stotra are described.

Chapter Six

This very brief chapter is called the Parapaddhati, and describes the general form of worship or puja for other devatas.

Chapter Seven

Describes a number of practices, pujas and mantras common to all, and makes quotations from Pancharatra texts, the Tantraraja and other sources. Various chakras used to determine the time of initiation, such as the Kulakula and the Rini-dani are described, as well as a number of mantras which are used to ensure that the root mantra, that is the mantra of the ishtadevata or desired form of divinity, is successful. These purifications relate to the birth, the life and other stages that a mantra is assumed to have.

Gandharva Tantra

Gandharva Tantra is an important work of the school of Shri Vidya and follows Vamakeshvari Tantra in having a left slant. This introduction to the Sanskrit text was written by M.S.Kaul in 1944 and is out of copyright. The abstract is of such interest to students and devotees Shri Vidya that it is well worth reproducing here. To preserve mantras and the like.

Abstract of Gandharva Tantra: Unlike the Agamas, the

Gandharva Tantra begins with the two stanzas, one of salutation to the Elephant-god and the other of benediction invoking the protection of the Goddess Kundalini. This fact in itself reveals that the Tantra must be more or less modern. The tradition regarding the appearance of the Tantra is that the rival sage Vishvamitra, being envious of the prophetic powers of Vasishtha, performs a difficult penance. Failing even thereby to obtain equality with Vasishtha he goes to the North and implores the help of Dattatreya who consoles him and reveals the Gandharva Tantra which he has heard from Nandikeshvara. The Tantra is in the form of a dialogue between Shiva and Parvati.

1st Chapter

It begins with the question from Parvati about Brahman, Yoga and the Body. Shiva defines the Brahman only. Thereupon Parvati thereupon enquires after some such secret lore as will enable the humanity to get release from the bondage of actions while enjoying the blessings of the earthly life. Shiva says that the Tantric lore is triune in nature as Tamasik, Rajasik, and Sattvik leading respectively to hell, heaven and emancipation. He advises that care should be taken to reject the first and informs that the subject matter of the Tantra is already revealed by Him to Krishna the son of Devaki and to Nandikeshvara. The former related it to Brahma and the latter to Pushpadanta and through him to Gandharvas. The sages received it from Brahma and Indra from Angiras and the king of the demons from Sukra. The result was that everyone became piously-minded including even Namuchi and other demons and the eternal foes of the latter, *i.e.* Indra and other gods lost their high positions. Brahma, pitying them went to Shiva and related the grievances of the Gods. Shiva, consequently revealed more or less the materialistic Shastras to deceive the demons. Among these are mentioned the Pashupata-shaiva, the Vaisheshika, the Nyaya, the Samkhya, the Charvakas, and the Bauddha. All the mantras also were polluted as it were with various defilements and the Tantras interpolated with doubtful passages.

2nd Chapter

In this Parvati requests Shiva to remove the doubts referred to in the first Patala regarding the Shakti lore. Shiva accedes to the request and relates the importance of the worship of Shakti as Turiya. She (Tripurasundari) is called Turiya (the fourth) in-as-much as Kameshvari is the first. Bhagamalini the second and Vajreshvari the third. After this Shiva reveals the Mantra called Panchadasi Vidya. The same is of three parts. The first is called Vagbhava, as it imparts mystery over all forms of speech. It is read as ka e ii la hriim. The second goes by the name of Kamaraja and is composed of the six letters ha sa ka ha la hriim. The third is named Shaktibija consisting of the four letters sa ka la-hriim. Panchadasi in this form is known by the name of Kamarajavidya. The forms of the same respectively worshipped by Lopamudra, Shiva and Shakti are sa ka la hriim, ha sa ka ha la hriim, sa ka la hriim, ha sa ka ha la hriim, sa ka la hriim; sa ha ka la hriim, ha sa ka ha la hriim, sa ka la hriim.

Then comes the Shodashi which is read as hriim ka e ii la hriim ha sa ka ha la hriim sa ka la hriim. The second form of the same is called Chintamani read as shriim ka e ii la hriim ha sa ka ha la hriim sa ka la hriim. The same Shodashi when read as hriim shriim ka e ii la hriim ha sa ka ha la hriim sa ka la hriim is called Saptadashi. Rajarajeshvari is read as shriim hriim kliim sauh Om hriim shriim ka e ii la hriim ha sa ka ha la hriim sa ka la hriim shriim hriim Om sauh aim kliim hriim shriim.

3rd Chapter

In the third Patala Shiva reveals the Panchami Vidya which consists of five Kutas. It is so called because the whole world consists of the five elements—Earth, Water, Fire, Air and Sky represented by the five Kutas respectively presided over by the Shaktis of Brahma, Vishnu, Rudra, Isvara and Sadashiva. These are Kameshvari, Vajreshvari, Bhagamalini, Tripurasundari and Para.

The Vidya is to be read as ka e ii la hriim, ha sa ka la hriim, ha ka ha la hriim, ka ha ya la hriim, ha ka la sa hriim. After this follows the description regarding the way in which the Sadhaka is to practise the Vidya and of the powers appearing as a result thereof. Ekadashaksari or the mantra of eleven letters comes last. It is read as ka la hriim, ka ha la hriim, sa ka la hriim.

4th Chapter

In this Patala is given the Kavacha of Rajarajeshvari called Trailokyamohana. The sage of the Kavacha is Shiva, metre Virat, and the Goddess Mahatripurasundari. The Kavacha is said to have various kinds of efficacy which are detailed in the text. It is efficacious not only when practised but also when borne on different parts of the body after being written on a birchbark. The way of writing is that the names of the practitioner and the object of practice are written in the triangle of the Shrichakra, the Mulamantra coming at the top and around it the alphabet and around the alphabet the Kavacha.

5th Chapter

In this Shiva describes the way in which the Shrichakra is to be drawn and the different deities are to receive their worship in the nine component parts of Shrichakra. The Shrichakra is to be drawn either on a metallic plate or the earth with red lead by the golden pen. Devi is said to have three forms, physical, mental and cognitive. The first is represented by Mudra, the second by Yantra and the third by Mantra. Mahatripura-Sundari as seated in the lap of Mahashiva is to be worshipped in the Binduchakra or the centre. The maidens forming the six limbs of Mahatripura-Sundari Sarvajna, etc. as also the Datal Eternities are to receive their worship in the four sub-quarters, centre and the quarters. The fifteen Kalas or fifteen vowels are to be worshipped in the triangle, five in the right side and five in the left side and five in the base of the triangle. The Datal Eternities are to be worshipped through the medium of the lady.

Attendant Deities: The three rectangles otherwise called Trailokyamohana chakra are respectively presided over by Brahma, Vishnu and Shiva. Lord Buddha, and the 10 Yoginis, Anima, etc. are to have their worship in the external rectangle; in the inner rectangle eight Mothers Brahmani to Mahalakshmi and in the innermost the Mudradevis Sanksobhini, etc. All the Deities of the three rectangles referred to above go by the name of Prakatayoginis. In the sixteen-petalled lotus called Sarvashapuraka Chakra, Brahma and sixteen Guptayoginis Kamakarshanika, etc. are to be worshipped. Eight Guptatara Yoginis and Shiva receive their worship in the eight-petalled lotus called Sarvasan-kshobhana. In the fourteen-sided figure called Sarvasaubhagyadayaka Chakra the sun-god and the Sampradayayoginis Sarvasankshobhini, etc. get their worship. Kulakaula Yoginis and Narayana are worshipped in the outer decagon called Sarvarthasadhaka and in the inner decagon Nigarbha Yoginis.

In the octagon called Sarvarogahara Chakra Rahasya Yogini Vasini, etc. receive their worship. The Divinities of the weapons of Paramashiva and Mahatripura-sundari receive their worship outside the triangle. Kameshwari, Vajreshvari and Bhagamalini which are the three Shaktis respectively of Rudra, Vishnu and Brahma are to be worshipped in the apex, left and right side of the triangle which is called Sarvasiddhiprada Chakra. These Shaktis are called Atirahasya Yoginis. In the centre of the triangle or the Bindu Chakra Paraparahasya Yogini Mahatripurasundari is to receive her worship. The worship of Mahatripurasundari is threefold, Para, Apara and Parapara. In the Apara the practitioner has to worship her as identical with the Kundalini in the body. In the Apara she is to receive her worship in Shrichakra with all the necessary articles of worship. The third form of worship includes both the inner and the outer aspects of the worship.

6th Chapter

In the sixth Patala Shiva, in reply to the query made by Parvati regarding the third form of worship, gives in detail instructions for the guidance of the practitioner. These are that the practitioner should rise early in the morning, sit on the Padmasana, have the Pranayama exercise and meditate upon his Guru seated in the thousand-petalled lotus, dressed in white, having two eyes and two hands with Vara and Abhaya Mudras, with his wife sitting on the left thigh and having two hands one holding the white lotus and the other engaged in the tight embrace of her husband. Mantra of the Guru is aim hriim shriim ha sa kha phrem, ha sa ksha ma la va ra yum, ha sa kha phrem, ha sa ksha ma la va ra im, hsauh and names of the Guru and his wife followed respectively by anandanatha padukam pujayami and ambapadukam pujayami. He should offer the objects of senses through the latter to his Guru, smell as perfume, sound as flower, touch as incense, form as lamp, flavour as Naivedya. After this comes the mention of Yoga which is defined as the unification of Jivatma and Paramatma and of the eight limbs of the Yoga and their detailed description. At the end is given the description of the centres of Kundalini in the body, *i.e.*, Muladhara, Svadhishthana, Manipura, Anahata, Vishuddha, Ajna and the thousand-petalled lotus above them, which in turn is followed by such instructions as those for bathing, application of collyrium to the eyes and the cleaning of teeth.

7th Chapter

In the above is given the way in which the devotee of Mahatripurasundari is to purify his body and perform the Sandhya. Bathing is said to be of three kinds as also the Sandhya. The kinds are physical, mental and psychic consisting in the purification respectively of the body, the mind and soul. In the three kinds of Sandhyas done in the morning, midday and evening the devotee is to meditate upon Mahatripurasundari in three forms.

The Gayatri of Mahatripurasundari is given as tripurasundari vidmahe kameshvari dhimahi Tannah klinnam prachodayat. Each of these twenty-four letters is said to have a peculiar colour and shape. The muttering of the same results in having release from different kinds of sins. Agni, etc. are the presiding deities of these twenty-four letters. Parashurama, Menanath, Agastya, Vasishtha and Dattatreya are declared to have freed themselves from heinous crimes such as matricide, etc. Every important event in the life of a practitioner is to commence with the worship of Tripurasundari and at the end of the Sandhya he is to offer a handful of water mixed with red flowers and the powder of Sandalwood to the sun under the name of Martandabhairava with the Mantra hriim hamsah martandabhairavaya prakashashaktisahitaya idamarghyam namo namah.

8th Chapter

The eighth chapter describes the sacrificial altar and the way in which the practitioner should, while entering it, bow to the presiding deities of the quarters and other deities such as Ganesha, Ksetrapala, etc. and how he should make an offering to the Bhutas with the Mantra Om hriim sarvavighnakrt sarvabhutebhyoh hum svaha, how he should remove all obstacles by muttering seven times Astra-mantra which reads as aim kliim sauh Astraya phat and how he should clean the sacrificial altar. Every article of worship is to be concealed till the invocation of the Devi. The Devotee is instructed to use his discretion in selecting Asana because the fruit of worship varies with the material of which the Asana is made. Red blanket is specially advised for the worship of Tripura. It must measure two cubits in length and a cubit and a half in width. He has to sit also on it under the instructions of Asana.

The posture peculiar to the worship of Tripura is Padmasana and the North is the quarter which the practitioner must face to secure success in the worship of Tripurasundari. The practitioner should take care that when worshipping he should have everything red, such as saffron ointment, red dress, red seat, etc. and while following instructions regarding the use of the articles of worship and their places, he should have the Shrichakra drawn in red lead, placed on the Simhasana. Towards his left he should place a stand with a pot full of water on a rectangular diagram. The stand is to be identified with the orb of the fire, the pot with the solar orb and the water with the moon and these must as such be worshipped with the Kalas of the fire, the sun and the moon. Anandabhairava and Anandahhairavi are to receive their worship through the pot. Conch and other vessels are also to be placed in particular places.

The purification of the five elements, constituting the body by removing the impurities attaching to them through the letters peculiar to the elements and by subsequent infusion of the energy of the Changeless Reality is to be attended to and performed as a preliminary to the worship. For this purification the practitioner is to deeply concentrate in the Bindu-chakra on the Turiya aspect of Mahatripurasundari. After this he has to protect himself against all sinister influences by Digbandhana with the Mantra Om hriim hamsah so-aham svaha aim kliim sauh tripurasundari mam raksha raksha.

9th Chapter

In the ninth chapter is given the Matrikanyasa. In performing this the practitioner is first to wash his hand in a particular manner with the mantra am aam sauh. Then he has to pay his attention to Rishinyasa, etc. Then comes the Matrikanyasa of which all details such as Rishi, etc. and the six limbs, etc. are given. Matrika is said to be of two kinds as Antarmatrika and Bahirmatrika. In case of the former letters of the alphabet are to be meditated upon in the six centres of the body Muladhara to Bhrumadhya. In the latter case letters of the Matrika preceded each by kliim are to be meditated upon, vowels individually in the sixteen different parts of the head, five groups of letters, gutturals to labials in hands, feet, sides, back, navel, and belly, identifiable with the seven constituents of the body, skin, etc. vital soul, individual soul and supreme soul in heart, armpits, throat, shoulders, lower end of the heart, hands and feet, belly and face. Different poses of the hands are also to be used in performing this Nyasa. Towards the end various results obtainable from Matrika-nyasa are described.

10th Chapter

In the 10th Patala the sixfold Nyasa of Shrividya as consisting of those of Ganesha, planets, constellations, Yoginis, signs of Zodiac and Pitha is described.

Ganesha Nyasa: In this the fiftyone Ganeshas, Vighneshvara to Ganesha with their Saktis are to be contemplated upon in the different parts of the body, forehead, etc. assigning each letter of the alphabet to a couple as am vighneshvaraaya shriyai namo lalaate.

Planets (Graha) Nyasa: The planets Sun to comet occupy the places heart, middle of the eyebrow, eyes, heart, throat, navel, anus and the feet. Their letters are respectively vowels, semi-vowels, gutturals, palatals, linguals, dentals, labials, aspirates and La & ksha.

Constellations (Nakshatra Nyasa)

Details regarding this are as given in the following table:

1	*2*	*3*	*4*	*5*	*6*	*7*	*8*	*9*	*10*	*11*	*12*	*13*	*14*
am aam	im	iim um uum	R.im R.im Lim LIm	em	aim	Om aum	kam	kham gam	dham ~Nam	cham	Cham jam	jham nam	Tam Tham
Fore-head	Right eye	Left eye	Right ear	Left ear	Right nostril	Left nostril	throat	Right should.	Left should.	Right elbow	Left elbow	Right wrist	Left wrist

15	*16*	*17*	*18*	*19*	*20*	*21*	*22*	*23*	*24*	*25*	*26*	*27*
Dam tham	Dham dhampam	tam ram	dam sham	nam sam	Bam lam	bham pham	Mam ham	yam am ah	lam	vam	Shamksham	Nam
Right hand	Left hand	navel	Right hip	Left hip	Right thigh	Left thigh	Right knee	Left knee	Right leg	Left leg	Right foot	Left foot

Yogini Nyasa:

The Nyasa of Yoginis is to be performed with the mantras in the following manner:

YoginiLakini	***Kakini***	***Shakini***	***Hakini***	***Yakini***			
Colour	White	Blue	Red	Yellow	Brown	White	All colours
Form	With one face of three eyes	With two faces having three eyes each	With three faces having three eyes each	With four faces having three eyes each	With five faces having three eyes each	With six faces having three eyes each	All faces
Emblem	Sword, wine-cup, trident, skull-crowned club	Trident, small hand drum, lotus	Thunderbolt, vara spare, abhaya	Trident, small hand drum, noose, abhaya	Bow, arrows, noose, goad, book, jnana-mudra	Small hand drum, rosary, skull, sword, book, jnana-mudra	All weapons
Constituents of the human body	Skin	Blood	Flesh	Fat	Bones	Marrow	Shukra
Seat	Throat	Heart	Navel	Linga	Muladhara	Middle of the eyebrows	Cavity in the head
Offering	Milk & rice	White food	Sugar & rice	Curd & rice	Beans & rice	Turmeric & rice	All edibles

The mantra of the second is kam kham gam gham ~Nam cham Cham jam jham ~nam Tam Tham raam riim ramalavarayuum raakinyai namo raakini as Rigdhaatugate dvaatrimshallakshakotiyoginisvaamini sarvasattvavasham kari aagyaam me dehi mama vicche The mantra of the third is Dam Dham Nam tam tham dam dham nam pam pham laam liim lamalavarayuum laakinyai namo laakini maamsadhaatugate Shodashalakshakoti-yoginisvaamini sarvasattvavashamkari aagyaam me dehi mama vicche.

The mantra of the fourth is bam bham mam yam, ram lam kaam kiim kamalavarayuum kaakinyai namah medodhaatugate ashtalakshakotiyoginisvaamini sarvasattva-vashamkari aagyaam me dehi mama vicche.

The mantra of the fifth is vam sham Sham sam shaam shiim shamalavarayuum shaakinyai namah shaakini asthidhaatugate chaturlakshakotiyoginisvaamini sarvasattva-vashamkari aagyaam me dehi mama vicche.

The mantra of the sixth is ham ksham haam hiim hamalavarayuum haakinyai namo haakini majjaadhaatugate dvilakshakotiyoginisvaamini sarvasattvavashamkari aagyaam me dehi mama vicche.

The mantra of the seventh is am aam im iim um uum R.im R.im Lim Lim em aim Om aum am ah kam kham gam gham ~Nam cham Cham jam jham ~nam Tam Tham Dam Dham Nam tam tham dam dham nam pam pham bam bham mam yam ram lam vam sham Sham sam ham ksham yaam yiim yamalavarayuum yaakinyai namo yaakini shukradhaatugate dashakoTiyoginisvaamini sarvasatt-vavashamkari aagyaam me dehi mama vicche.

Rashi Nyasa: Like the two previous Nyasas, the above Nyasa also is prescribed in the table as given below.

Pitha Nyasa: In this the practitioner has to identify the fiftyone Pithas Kamarupa, etc. with the different parts of the body and in doing so the letters of the Matrika are to go along with the names of the pithas one with each with a Bindu at the top. The chapter ends with the spiritual powers that the practitioner of this Shodhanyasa gets.

	1	2	3	4	5	6	7	8	9	10	11	12
Letters	a.m aa.m i.m I.m	u.m U.m R^i.m	R^I.m L^i.m L^i.m	e.m ai.m	o.m au.m	a.m aH sha.m Sha.m sa.m ha.m	ka.m kha.m ga.m gha.m ~Na.m	cha.m Cha.m ja.m jha.m ~na.m	Ta.m Tha.m Da.m Dha.m Na.m	ta.m tha.m da.m dha.m na.m	pa.m pha.m ba.m bha.m ma.m	ya.m ra.m la.m va.m ksha.m
Seat	Right ankle	Right knee	Right testicle	Belly right	Right shoulder	Right head	Left head	Left shoulder	Left belly	Left testicle	Left knee	Left ankle

11th Chapter

In the 11th Patala are described first the Nyasa of (1) Asana, (2) Vasinyadivagdevata, (3) Mulanga, (4) Navayoni, (5) Chaturvyuha, (6) Tattvanyasa, (7) Mulavidya, and (8) Sammohana.

In the first, Asanas, *i.e.* Amritarnavasana, Potambujasana, Atmasana, Chakrasana, Sarvamantrasana, Sadhyasana, Sadhyasiddhasana, Paryankashaktipithasana, and Mahapretasana respectively presided over by Tripura, Tripureshvari, Tripurasundari, Tripuravasini, Tripurashri, Tripuramalini, Tripurasiddha, Tripuramba, and Mahatripurabhairavi are to be contemplated upon in feet, knees, thighs, hips, private part, Muladhara, navel, heart and cavity in the head.

The Mantras to be used are: (1) am aam sauh tripuraamritaarnavaasanaaya namah (2) aim kliim sauh tripureshvariipotaa, nujaasanaaya namah (3) hriim kliim sauh tripurasundaryaatmaasanaaya namah (4) aim hvliim hsauh tripuravaasiniichakraasanaaya namah (5) hsaim hskliim hssauh tripurashriisarvamantraasanaaya namah (6) hriim kliim blem tripuramaa-liniisaadhyaasanaaya namah (7) hriim shriim sauh shivamahaapretapadmaasanaaya namah.

In the second, eight goddesses of speech are to be meditated upon as presiding deities of (1) Vowels, (2) Gutturals, (3) Palatals. (4) Linguals, (5) Dentals, (6) Labials, (7) Semi-vowels, and (8) Aspirates. Their names, seats and Kutas are (1) Vashini, cavity of the head, rbluum (2) Kameshvari, forehead, klhriim (3)`Modini, middle or the eyebrows, nvliim (4) Vimala, throat yluum (5) Aruna, heart, jmriim (6) Jayini, navel, hasalavayuum (7) Sarveshvari, Linga, jhamarayuum (8) Kaulini, below the Linga, kshmriim

In the third, the six limbs, *i.e.* heart, head, tuft of hair on the head, armour, eyes and the weapons are to be located in the five fingers of the hands and their palm and back. While doing so the practitioner is to mutter in each case the mantras of Shrividya twice. The six poses of the hands are prescribed for these six limbs, the three fingers—middle, ring and index for Hridaya; two fingers—mid and index for Shiras; thumb for Shikha; ten fingers for Kavacha, the three above referred to for eyes and the two above mentioned for Astra.

In the fourth, Vashini, etc. are to be meditated upon in the different parts of the body named in book and the muttering of all the three Kutas of Shrividya is to be done in each case.

In the fifth, the four forms of Mahatripurasundari, *i.e.* Kameshvari, Vajreshvari, Bhagamalini and Mahatripurasundari are to be contemplated upon below the Linga, in the heart, in the middle of the eyebrows and in the cavity in the head with the four famous centres of her worship and their priests.

The mantras to be muttered are (1) aim agnichakre kaamagiripiithe mitreshanaaathaatmake kaameshvariirudraatmashakti shriipaadukaayai namah (2) kliim suuryachakre jaalandharapiithe Shashthiisha-naathaatmakam vajreshvariivishnavaatmashakti shriipaadukaayai namah (3) sauh somachakre puurnagiripiithe uddiishanaathaatmakam bhagamaalinii-brahmaatmash-aktishriipaadukaayai namah (4) turyabiijam brahmachakre uduyaanapiithe charyaanaathaatmake parabrahmaatma-shaktishriipa-adukaayai namah.

In the sixth, the three Tattvas Atma, Vidya and Shiva are to be located in the three parts of the body, *viz.* (1) from feet to the place below the Linga, (2) thence to heart, (3) from the heart to the middle of the eyebrows, uttering the following mantras:

1. aim aatmatattvavyaapikaayai mahaatripurasundaryai namah.
2. kliim vidyaatattvavyaapikaayai mahaatripurasundaryai namah.

3. sauh shivatattvavyaapikaayai mahaatripurasundaryai namah. In the seventh, the three Kutas of Shrividya are to be located in hands, etc. individually and the fourth bija is to have its place in the cavity in the head. The fifteen syllables of the Panchadasi are located in the cavity in the head, Muladhara, heart, eyes, ears, mouth, arms, back, knees and navel.

In the eighth, the practitioner is to consider himself as endowed with a body of mantras and meditate upon Shrividya in the mantra form. While this is being done, Yonimudra is to be placed in the cavity of the head, forehead, middle of the eyebrows, face and heart. What has been described above relates to the external worship. As regards the internal, the practitioner is first to purify his mind and soul through pranayama which is said to be of the highest value in the spiritual domain. By Prana is meant the life breath and by Ayama the control thereof. Pranayama consists of the three functions—inhalation, exhalation and retention.

Awakening of Kundalini. While doing the pranayama referred to above, the practitioner is to mutter mentally either the whole mantra or its initial syllable. The Kundalini lying dormant in the Muladhara is to be contemplated upon as a thin red flame rising upwards when awakened with the help of Aumkara and piercing the six life centres in its upward march, till it reaches Shiva and becomes united with Him in bliss. Thence she comes down again through the same way to Muladhara.

The Kundalini while moving up from Muladhara to the heart receives the name of vahnikundalinii being fiery, from heart to the throat suuryakundalinii being solar and from throat to head somakundalinii being lunar. The three Kutas respectively get assimilated to her in the three places referred to above. In the Bindu, the fourth form of Kundalini as embracing all the Bijas and all forms of light is to be contemplated upon.

Meditation on Kundalini, which constitutes the inner worship, helps the practitioner to acquire freedom from all sins.

12th Chapter

In the 12th Patala the mental offering of worship and the mental approach to and unification with Mahatripurasundari are at length described. In the first, the practitioner is to sit on the prescribed Asana keeping his body erect and with the hands full of red flowers placed near the heart, close the eyes and feeling as it were the august presence of Mahatripurasundari in the heart, offer worship to her with the mental 16 requisites of worship. The Yogis and sages have recourse to this worship only.

In the second, *i.e.*, the Dhyanayoga, the practitioner has to select a solitary place and seat himself in a particular posture with all his mental operations brought under control. He should feel oneness with the Transcendental Self through absolute introspection. In order to rise to that plane of consciousness, he has to immerge the different elements of which the world is composed in the causeless cause. The order in which these evolutes are absorbed from the following into the preceding is given as follows:

The earth element merges in the water, water in the fire, fire in the air, air in the ether, ether in the mind, mind in the ego, ego in the manifest, the manifest matter in the unmanifest, and the unmanifest in the Supreme Self.

After the absorptive meditation is over, he has to think of evolution. In this he is to think of evolution of the world from the causeless cause through matter, ego, mind, etc. down to the frog, Kalagnirudra, Adharashakti in the form of a crocodile and boar holding on the tooth the earth. There he has to think of the nectar-ocean, the coral island, the golden hill, the heavenly garden Kalpodyana and of streams and lakes. On the shore of the highest among the lakes he should imagine the existence of the desire-granting tree with a jewel-pavilion having four gates fitted with diamond doors and coral thresholds and other decorative parts.

After having done so, he is to imagine the presence of one big lotus enclosed in another placed in the centre of the Simhasana arranged on a platform of jewels inside the pavilion described above. The outer lotus represents all the elements, its bulb the bliss, its stalk the consciousness, all the evolutes of Prakriti furnishing the thirtytwo leaves and the fifty letters furnishing the pericarp of the lotus. Inside the lotus he has to think of the sun, moon and fire one placed below another. In the fiery orb there stands the triangular mass of light borne by the five karanas.

In the centre of this triangle stands, the third lotus in which the practitioner is to think of the seven goddesses Tripura to Tripuramba. Of the last, beautiful description of each part of the body is given in detail in the text. While thinking so, the practitioner is to be at one with Mahatripurasundari.

13th Chapter

In this, stress is laid first of all on the actual realization of the unity between the worshipper and the worshipped. Everything else used in the worship receives sanctity by it. A devotee is enjoined to worship Shrividya both mentally as will as physically in case he happens to be rich enough. All vessels required for worship may be of copper. A devotee may, if he can, make five vessels from five jewels which are named as Garuda, Manikya, Vajra, Vaidurya and Nila. The Bijas of the five jewels are gluum sluum pluum mluum nluum The pot which is to contain Vishesarghya is to get a place in the ceremonious manner between Shrichakra and the worshipper on a diagram drawn on the earth. The diagram is to consist of a rectangle, a hexagon, a circle and a triangle, one enclosing the other. The mantra to be used is aim kliim sauh mahaatripurasundaryaa arghyapaatraasanam saadhyaami.

The stand of the pot, the pot and the liquid poured into the pot represent the fire, the sun and the moon and hence the former are to receive the worship respectively of the Kalas of the latter. The liquid to be poured is to be one of the following: wine, milk, coconut-juice, sugarcane juice, sugar, honey, ghee. Presence of the sacred waters is to be invoked in the pot from the solar orb with the Ankushamudra. Then the pot is to be covered over with the hands in the Matsyamudra. The jewels Manikya etc., are to be put into the liquid of the pot, the goddess of jewels being worshipped through the jewels. Her mantra is nine-syllabled as shriim hriim gluum sluum pluum mluum nluum hriim shriim.

The Matrika is to be contemplated upon in the form of a triangle, vowels occupying the base, consonants ka to ta and tha to sa the arms and ha and ksha the centre. The three angles and the centre are to be identified with Kamarupa, Jalandhara, Purnagiri and Uddyana presided over by the four forms of Tripura. Ananda-bhairava described as ten-armed, five-faced and having ten emblems and Suradevi similarly formed in his lap are to be worshipped inside the pot. Amritikarana, Sakalikarana and Paramikarana are to be done afterwards with Dhenumudra and Mahamudra as also with Musalamudra, Yonimudra and Galinimudra. The mantras of these are (1) hasakshamalavarayuum aanandabhairavaaya vaushath (2) sahakshamalavarayuum suraadevyai vaushath.

In the purification the verses to be mattered are (1) hamsah etc., (2) pratidvishnu, etc. (3) tryambakam yajaamahe, etc. (4) tadvishnoh, etc. (5) vishnoryoni, etc. The mantra aim kliim, etc. is to be muttered ten times. The last items of purification are to be done with the help of the Sandipini Vidya aim vada vada, etc. At the end, the practitioner is to offer libation to Shrividya from the liquid thus consecrated.

14th Chapter

The 14th Patala deals with the worshipping of Kameshvari. In it the practitioner is to offer the Asana on the Kamapitha to Kameshvari invoking the presence of Adharashakti and others upto Kameshvara who serve as her seat. Kameshvara is to be contemplated upon as four-armed, seated on the white lotus and holding in the left hand trident and bow and in the right white lotus and Bijapura.

Then Kameshvari is to be contemplated upon as being seated in the lap of Kameshvara in response to the invocation with the mantra aim hriim sauh mahaatripurasundaryamritachaitanyamuurtim kalpayaami. The rites of removing the sinister influences and Digbandhana come after this. The vitalizing of the idol is to be conducted by the Sadhaka, after holding the breath and realizing through contemplation the presence of Mahatripurasundari in the heart, by conceptualizing the vitality in the handful of flowers after it comes out through the nostrils with the Bija yam

Trikhanda and Avahani Mudras, which are described in the text, should be used as also the Avahanividya hrsaim hskliim hssrauh invoking the presence of Mahatripurasundari in the idol.

This being over, the practitioner has to address his prayers to Mahatripurasundari for firm stay, for being eye to eye and mind to mind with the practitioner, for success in completion, for being invisible to the undeserving, for being blissful and finally for being merciful using the Mudras, such as are prescribed for these acts.

After these preliminary rituals, comes the offering of three handfuls of flowers and three libations of the consecrated liquid with the mantras (1) aim kliim sauh mahaatripurasundarii shriipaadukaam puujayaami (2) aim kliim sauh mahaatripurasundarii shriipaadukaam tarpayaami.

Exhibition of the nine Mudras Samksobhini to Yoni follows the offering of the following prayer to the Goddess:-

> *"O Goddess: May you be kind to me, may you accept my worship as I, though poor physically as well as in material resources, offer it in a devotional spirit."*

After the exhibition of the Mudras, the Sadhaka is to present the seven items of offerings using namah svadhaa vaushat wherever necessary.

Tarpana is to be done with the Tattva Mudra in which the thumb and the index fingers of the left hand are to be joined. At the end, come Japa and offer of the sacrifice, ointment, collyrium, mirror and umbrella.

15th Chapter

The 15th chapter describes the things to be offered in worship. These are Asana, Padya, Arghya, Achamana, Madhuparka, Snana, Vastra, Bhushana, Gandha, Anjana, Pushpa, Dhupa, Dipa, and Naivedya.

Asana or the Seat: It is to be either of gold or silver or cloth or jewels or flowers or wood. Among the cloths red blanket is preferable.

Padya means Water for Footwash: Cold water mixed with sandal-wood paste and Agaru is to be offered either in a pot of gold or silver or in a conch.

Arghya for the Head: This is to consist of white mustard beans, sesamum, milk, red sandal and red flowers mixed in water. This is to be offered in a conch like other articles with a particular mantra given in the text.

Achamana or water for sipping. This is to be pure and scented with camphor and Agaru and offered in the same pot as is prescribed for Padya.

Madhuparka: This consists of curd, ghee, water, honey, sugar and sugarcane juice or of cocoanut-juice, sugar, curd and ghee in equal proportions and honey in a larger quantity and offered in a bronze or silver pot.

Snana: Water mixed with camphor, Agaru, musk, Gorochana, saffron, cocoanut-juice, honey, sugarcane-juice, sugar, five products of cow, particular herbs called Sarvausadhi, is to be offered in a jug or conch for bath.

Vastra: A pair of cloths figured and silken is to be presented for dress.

Bhushana: Sixteen kinds of ornaments detailed in the text all made of gold and set with jewels are to serve as Bhushana.

Gandha: Five kinds of perfumery obtainable from certain fragrant substances by powdering, friction, burning, pressing, and from muskdeer are also to be offered.

Anjana: This is collyrium for application to the eyes. It is to be made of lampblack collected in a golden pot from a burning lamp having ghee or oil and scented with camphor.

Pushpa: Flowers to be offered are to be fresh. Certain species such as Bakula, Kesara, etc. are declared dear to Tripura. Garlands also made of these flowers please her very much.

Dhupa, or the Incense: It is to be offered in a pot and consists of sixteen substances, Patra to Agara given in the text.

Dipa: Lamps preferably of ghee with a wick of cotton are offered for illumination. These should be of gold or silver or earth or iron and placed on stands of similar material with the shape of a tree.

Naivedya, or Food. It comes last and is to be presented and placed either on the right or on the left or in front, lamps coming always either on the right or in front, incense on the left or in front, scents, flowers and ornaments always in front.

16th Chapter

Sixteenth Patala is devoted to the description of the offering, such as, drink and food and of the way and varieties of salutation. Among the first comes wine. This is to be distilled in the following manner:-

Well filtered sugarcane-juice should be allowed to grow stale for some time. Mango juice and honey should be mixed with it as also juices of jujuba, rose-apple fruit, dates, cocoanut, wine and intoxicative powders such as of mohini, jati, datura, jatiphala, mace, cardamom, black and long pepper, dry ginger and Triphala. The same should be scented with camphor and kept in a gold or silver or glass vessel and mixed with sugar and allowed to remain corked, being covered over with red cloth for seven nights. When at the end it becomes thus prepared and ready for use, it should be scented again with sandal paste, Aguru and camphor. This kind of wine is reserved for the more or less abstemious people and for those otherwise—minded a common species of the same is prescribed. Its use is admissible only in a strictly ceremonious way.

The four kinds of food detailed as chewable, suckable, lickable and drinkable come after drink. Every variety of solid food is to be offered when properly cooked and spiced. Every sweet fruit also is to be offered in cups of gold or silver and food in dishes or gold, silver, copper or stone or even in lotus leaves. The mantra to be muttered in this offering is mahaatripurasundari vaivedyam grihaana. After offering the perfumed water for sipping come the offerings to five lifebreaths

Prana, etc., with the five fingers. Offering of dessert comes last except Namaskara.

The latter is said to be of three varieties as physical, lingual and mental. The physical again is of three kinds as that in which hands are folded and placed on the head and as that in which the head only touches the earth. The Lingual also is of three kinds. First is that in which self-composed prayers are offered the second in which prayers from the Puranas, Vedas or the Tantras are offered and the third in which prayers composed by others are offered. The mental also has three kinds according as the mind thinks well, ill or indifferent.

Pradakshina or circumambulation is included in the bowing. The former is of six varieties as triangular, hexagonal, semicircular, circular, staff-like with eight limbs and the terrific. Moving from the south to the northwest, thence to the northeast and again thence to the south goes by the name of Trikona. That from the middle of the south to the northwest, thence to the northeast and thence to the middle of the south and again from the southeast to the southwest, thence to the middle of north, and thence to the southeast, goes by the name of Shatkona.

Moving from the middle of the south to the northwest and thence turning round and going to the middle of the south, goes by the name of Ardhachandra. Moving with the right hand extended and the head bent from the middle point in front of the Deity, around from the right is calied Pradaksina.

Rising from one's own seat, going back several steps and then prostrating on the earth is called Danda. Bowing in which the earth is touched with chin, mouth, nose, jaws, eyes, ears, and cavity in head goes by the name of Ashtanga. Touching the earth with the skull goes by the name of Ugra.

17th Chapter

In the 17th Patala worship, in the nine component parts of Shrichakra, of the attendant deities classified under nine categories technically called Prakatayoginis, etc., is given along with the description of the Mudras and their use. In the first part called Trailokyamohana represented by the three rectangles the worship is to be conducted from the right unlike other parts in which the left side is to be adhered to. In the external rectangle Buddha as the great Lord of Yoginis is to be worshipped with the Mantra bum buddhaaya namah along with the ten psychic powers Anima., etc. The first four powers occupy the quarters beginning with west, second four the sub-quarters beginning with the northwest and the last two come below and above. The mantra to be used in the worship of a psychic power is hriim shriim and the name of the psychic power and shriipaadukaam puujayaami. In the inner rectangle the eight mothers Brahmani, etc. and the eight Bhairavas Asitanga, etc. are to be adored. In the innermost rectangle the Mudra Devis Sarvasanksobhini, etc. receive their worship. For the description of Sarvasanksobhini Mudra see the text. The Bija of the Mudra is draam.

In the second part, *i.e.* sixteen-petalled lotus bearing the name of Sarvashapuraka sixteen datal eternities, and Brahma receive their worship. The Mantra of Brahma and the Bija of Dravini Mudra to be shown here are respectively Om brahmane namah and driim.

In the third part, *i.e.* eight-petalled lotus called Sarvasanksobhana, Shiva and Uma, the eight Guptatara Yoginis Anangakusuma, etc. are adored. Mantras of Shiva and Uma, eight Guptatara Yoginis and the Bija of the Mudra are kliim shivaaya namah, hriim umaayai namah, Gutturals, Palatals, linguals, dentals, labials, semi-vowels, aspirates, La ksha and kliim.

In the fourth, *i.e.* fourteen sided figure called Sarvasaubhagyadayaka the sun-god, the Sampradaya-yoginis Sarvasankshobhini, etc. and the Mudra called Sarvavashakari come in. The Bija of the Mudra is bluum.

In the fifth, *i.e.*, external decagon called Sarvarthasadhaka Vishnu (as Hrishikesha), Kulakaula-yoginis and Sarvonmadakari Mudra find their worship. The mantra of Vishnu and the Bija of the Mudra are kliim hrishiikeshaaya namah and sah.

In the sixth part, *i.e.* inner decagon called Sarvarakshakara, Vishnu (Trailokyamohana), Nigarbhayoginis, Sarvajna, etc. and the Mahankusha Mudra receive their worship. The Bija of Vishnu and the Mudra are kliim and krom.

In the seventh part, *i.e.* octagon called Sarvarogahara, Rahasya-yogints Vashini, etc. receive their worship. The Mudra to be exhibited is khechari of which the Bija is hasakhaphrem.

In the eighth part, *i.e.* the triangle called Sarvasiddhiprada Kameshvari, Vajreshvari and Bhagamalini, the three shaktis respectively of Rudra, Vishnu and Brahma are to be worshipped in the apex, left and right sides of the triangle. These Shaktis are called Atirahasya-yoginis. Mudra to be exhibited there is Bija-mudra, of which the Bija is hsauh.

In the ninth part, *i.e.* centre of the triangle called Sarvanandamaya Parapara-rahasya-yogini, *i.e.* Tripura-bhairavi is to be worshipped along with Mokshasiddhi and the lion the vehicle of the Goddess. The Mudra to be shown is Yoni-mudra of which the Bija is aim. The deity-in-chief is to be worshipped in all the eight previous parts under different names—Tripura, Tripureshvari, Triparasundari, Tripura-vasini, Tripurashri, Tripuramalini, Tripurasiddha and Tripuramba.

18th Chapter

The 18th chapter gives a detailed description of Japa. It is said to consist in the repeated muttering of certain syllables. Meditation on and muttering of the Mantra are prescribed to be successive, both being necessary for acquiring psychic powers. The rosary meant for Japa is to be considered as consisting of the letters of the alphabet in both the orders, successive and reverse as also of their eight groups.

It therefore contains 108 beads. In the absence of the rosary a devotee is to keep an account of his muttering by using finger-joints. Top and mid-joints of the index fingers are to be avoided. The rosary made of crystal, etc. before being used for Japa is to be placed either in front or on the left hand and then to be sanctified with proper worship after being besprinkled with water. The mantra of Mala is aim maam maale mahaamaale sarvashaktisvaruupini chaturvargastvayi nyastastasvmaath me siddhidaa bhava.

After being so purified in the said manner, it is to be held in the right hand with the mantra glaum avighnam kuru maale tvamh Care is to be taken in rotating it in a cautious manner, so that it may not slip down from the hand. The rosary along with Yantra, mantra and mudra is strictly prohibited from being shown to any body. The Index finger is not used at all in the rotation of the rosary.

The five fingers, thumb to the little finger are to be used in the Japa respectively meant for emancipation, destruction of the enemies, earning of wealth, pacification and attraction. Virasana is helpful to Japa. Japa is of three kinds loud, low and mental, the loud being inferior to the low, and the latter to the mental.

It is to be done neither too quickly nor too slowly. The mind at the time of Japa should be entirely withdrawn from the sense-objects. Unlike a hymn, Japa is always to be mental in the case of Shrividya. The minimum number of times that Japa is to be done comes to 108, the middle to a thousand and the maximum to ten thousand. Rules to be observed in Japa are that the practitioner is lightly-dressed, must not have his hands wrapped up with cloth and must have his head naked and free from cares, be neither angry nor hasty, neither diseased nor hungry. He must above all have his mind concentrated.

Japa at the end is dedicated to Shrividya. Pranava (Omkara) as forming the essence of the Vedas is to precede and follow. Japa is always to be accompanied by Homa (sacrifice). Offerings in the Homa should be equal to the 10th part of the mutterings of Japa. In case a practitioner be not in a position to do Homa he is advised to mutter the Mulamantra twice the number of the offerings of Homa.

Worship offered to the Deity with the various prescribed articles helps the practitioner to get desirable objects determined in each case by the former. The latter are described at length in the text. The nine Mudras are to be exhibited at the beginning and the end of worship.

The following things lead to success in worship:

Solitude, Spiritual purity, abundant purity of the worshipping material, presence of sweet-smelling substances, mental fortitude, affablehess of manners, nobility of disposition, beautiful dress, devotional tone, etc.

19th Chapter

In the 19th chapter worship and offerings to Vatuka, Yoginis, Kshetrapala and Ganesha are described. The latter are to be located in the four diagrams drawn in northeast, southeast, southwest and northwest of the seat of Shrichakra. Offerings are to consist of various edible (kind of meat), fish, sugar, curd, milk and flour.

Vatuka is to be meditated upon as dwarfish, lotus-eyed, gold-yellow and four-armed bearing in the hands mace, conch, lotus and disc. The mantras of worship and offering to Vatuka with the left hand thumb and ring-finger are blom batukaaya nama ehyehi deviiputra batukanaatha kapilajataa-bhaarabhaasvara trinetra jvaalaamukha sarvavighnaanh naashaya naashaya sarvopachaarasahitam balim grhna grhna svaahaa.

Yoginis: These are of optional bodily form, gold-yellow, wearing the garland of skeletons, dressed in red and bearing in arms linga, noose, skull and goad. Mantra of offering with the left hand thumb, mid-finger and ring-finger is yaam yoginibhyah sarvavarnamyoginibhyo hum phath svaahaa.

Kshetrapala: He is described as collyrium-black and bearing sword, skull, trident and small hand-drum. Mantra of the offering with the left-hand fisted and having the index-finger stretched is kshaam kshiim kshuum kshaim kshaum kshah kshetrapaala dhuupadiipaadisahitam balim grhna grhna svaahaa.

Ganesha: He is described as elephant-headed, four-armed, naked, big bellied, holding noose, goad and skull-bowl filled with wine, with shakti in the lap. Mantra of the offering with the left hand fisted having the mid finger raised up is gaam giim guum ganapataye varavarada sarvajanam me vashamaanaya balim grhna grhna svaahaa.

After making offerings to various kinds of beings the practitioner is to offer himself with the mantra itahpuurva, etc. and the prayers imploring the goddess to accept the offerings overlooking the errors of commission and omission. Towards the end of the chapter it is described how the practitioner has first to immerge the attendant deities in the goddess and then the goddess herself with the help of Pranayam̃a withdrawn from the outside world in the lotus of his own heart. Shoshika, the goddess that is entitled to the Nirmalya receives the practitioner's attention at the end.

20th Chapter

In this Shoshika is described at length. Her Mantra is given in two forms as follows (a) uchchishta, etc. (b) aim hriim, etc. She is described as seated on a corpse, dressed in red, bearing in two arms skull-bowl and scissors. The practitioner of her mantra is to make offerings to her and soon after taking food and without washing his body mutter her mantra. Her Homa and Tarpana are to be performed in a rectangle drawn on the ground with curd, white mustard and rice. By offering various kinds of meat in the fire in her name, the practitioner acquires certain desired objects.

21st Chapter

In the above the last rites of worship are described. The practitioner is first of all advised to offer Nirmalya to Ucchistamatangi or Soshika taking care that no drop falls down on the earth. The water in the conch is to be held in the left hand and after sanctifying the same with mutterings of the mantra the body is to be besprinkled with the same water through the right hand. Washings of the feet of the goddess are to be drunk and the remainder is to be used in rubbing the head and the rest of the body. Arghya is offered to the Sun-god with his mantra for completing the ceremony. In this, the knees of the worshipper are to touch the ground and the mind to be concentrated. Salutation to and seeing of the sun are to begin and end the ceremony. The place where the vessels have been arranged should be duly washed with the water and the washing should be thrown into the water or near the root of a tree.

After offering the food to the preceptor and his wife or in their absence to the girls, the remaining part of the same should be eaten by thedevotee. Such a food is considered as Amrita. Sipping water before and after is necessary as also reciting the following prayer:-shivo daataa shivo bhoktaa, etc.

Ladies and Brahmins are to be fed each time the worship is performed. Offering of food to the preceptor, girls, Brahmins and Cow helps to remove the sin of errors committed in worshipping. In doing Japa the concentration of the mind is to be aimed at by believing in and feeling identity with Shrividya either as transcendent or as immanent.

22nd Chapter

This chapter describes the duties of a Sadhaka. These are:- (1) Realization of the Supreme Reality as identical with the individual soul. Anything offered with this conviction receives sacredness. (2) The worship of Tripura. It is to be done either as Nitya or as Naimittika or as Kamya. By the first is meant that which is done everyday and omission of which is sinful. By the second is meant that which is done occasionally with a definite end in view and failure in the performance of which is artended with the evil consequences. The last denotes that kind of worship which is done avowedly to gain some object. In rendering all these kinds of worships it is to be kept in view that nothing is done contrary to the popular usage. Worship in all its forms if done without any desire to gain this or that object is viewed as Sattvika. (3) Performance of the acts that are prescribed for him by the Shruti and Smriti such as service to the parents and entertainment of friends and relations, maintenance of protection of the dependents; preservation of peace and order with a view to honouring the righteous and punishing the wicked. (4) Meditation upon Tripura Sundari and Tripura Bhairava four times a day in the morning, mid-day, evening and midnight.

23rd Chapter

This describes the importance of worshipping the girls, their prescribed ages and the way in which they are to be worshipped. Maidens also are mentioned in this chapter to receive the worship of the practitioner as representing the eight mothers Brahmi, etc.

24th Chapter

It describes the importance and fruit of offering worship to the girls and the Prayaschitta or penance, that is to be done by the devotee when he misses performance of daily worship. It lays stress on the daily worship as compared with the occasional or purposeful and instructs that on the 14th, 8th or 15th days of the bright half of the month as also on the 14th day of the dark half falling on Tuesday, Yajna or sacrifice should be offered to Tripura. Yajna is considered as a body in which meditation, worship, muttering of the mantra and Homa constitute the four hands, devotion head, faith heart, skill in action eyes, knowledge of the Atman the Soul and Nyasas the remaining parts of the body. Such a 14th day of the dark half of the month on which Tuesday and the constellation Tishya fall simultaneously is the most sacred day for the devotee of Tripura.

25th Chapter

This chapter is devoted to the description of the sacred spots in India and the merits that a devotee gains by offering worship in these places. The sacred places are river banks, caves, mountain tops, sources of water, confluences of rivers, forests, gardens, mountain feet, Tulasi groves, cow-sheds, root of the Ashvattha or Amalaki, temple of Shiva, sea shore of ocean, solitary places and those where the preceptor lives or which help in the concentration of the mind, the cities of Benares and Dwarika, the Vindhya mountain, Aryavarta, Brahmavarta, Prayaga, Pushkara, the river Karatoya, Nandikunda, etc. It also tells that the devotee is not held back from worship by the consideration of impurity arising from death or birth and that he should offer worship mentally in times of such a calamity.

26th Chapter

It deals with Diksha or initiation. It is necessary for the practitioner of mantras because it imparts right knowledge and removes sins. Before having been done by man to man initiation was started by the holy line of nine divine teachers, *viz.* Prakasha, Vimarsha, Ananda, Jnana, Satya, Purna, Svabhava, Pratibha, Subhaga, the first three living eternally in heaven, the middle three partly in heaven and partly on earth and the last three only on earth.

These receive their worship between Vimala and Jayani in Shrichakra. Qualifications are laid down of both the initiator and the initiate. The former among other things must have all good qualities = be free from vanity, greedless, healthy, pure, speaking truth and knowing sacred laws of the Tantras. The latter must be orthodox, merciful, attached to the teacher, considerate and greedless. He should avoid all acts of indecency, especially when in the presence of the teach and should have as much faith in him as in his own God or Mantra and be respectful towards him. The months which are prescribed for initiation are given in the following order of importance:-

Assoon, Katak, Baishakh, Phagan, Maghar, Chet, Har and Magh. Days of the lunar reckoning and the week days prescribed for the said purpose are 15th, 5th, 2nd, 7th, and 13th and 10th and Sunday, Thursday, Friday, Monday and Wednesday.

The initiation is said to be of three kinds that of mantras, that of Shakti and that of Shiva. The first involves the use of mantras, worship, postures, meditation and concentration, etc. The second has a reference to that in which the initiator infuses his own spiritual power in the initiate. The third is that in which a spiritualist par excellence sanctifies the person with no regard to qualifications merely by his glance to the extent that the Reality becomes visible to him.

The end of the chapter explains the details of initiation such as selection and preparation of a place for initiation, making offering to Vastupala together with the Vedic recitation and blessing of the learned Brahmins, sowing of seeds, and rendering worship to the Dikpalas with a view to invoking their help in performing initiation, purification of the place, Shrichakra and the disciple with Panchagavya, performing Homa and disclosing the mantra called Svapnamanava, *i.e.* hili hili shuulapaanaye svaahaa to the disciple and putting him to sleep afterwards, The disciple is instructed to carefully remember the dreams good or bad he has seen in sleep.

The good dreams are those in which the objects seen are a girl, an umbrella, a chariot a lamp, a palace, a river, saffron, etc. Bad dreams on the other hand, involve reference to an ass, a crow, ditch, etc. The preceptor has to make a hundred offering with the Astra mantra in case the initiate sees bad dreams.

27th Chapter

This also is devoted to the description of initiation. The seeker after initiation is to approach the initiator with the request for initiation after having duly done worship to gods and the manes. Initiator is to enter the pavilion raised for the purpose and instal a water-pot of gold or silver or copper on the Shrichakra drawn on the earth. The water-pot is to be filled with fragrant substances and certain herbs and varieties of earth, and wrapped with a couple of cloths. The same is to be placed on a stand.

The stand, the water-pot and the water respectively representing the Fire, the Sun and the Moon are to receive worship through their Kalas. The candidate for initiation is asked after this to fill his hands with flowers and throw them so as to see whether their throw is in the auspicious direction or otherwise. In the latter case the initiator is to ward off the evil by performing a Homa. After performing worship to the attendant deities of Tripurasundari sacrificial food is to be prepared of which a part is to be offered to her and the rest to the fire, care being taken that the fire is ablaze.

The candidate is to have his face covered with a cloth and then be besprinkled with the water from Vishesharghya and the initiator is to request the goddess to be kind to the candidate and inspire him with her devotion. After this he has to receive instruction from the initiator as to the line of action he should follow in future.

The initiator should imagine that the fetters of sins binding the body of the candidate have been burnt down with the fire of Kundalini rising from the Muladhara to the Brahmarandhra and also enter his body through the Yogic power and unite his soul with his own.

After having him seated on a diagram of the alphabet the onitiator should bathe him with the water of the water-pot, muttering the three Bijas of the Mulamantra and then dress him and joint him with sandal-wood and other pastes. He should also instruct the candidate to perform the Matrika-nyasa and the Mantra-nyasa at the end of which his face is to be unveiled and he sees things required for the worship and takes food, while muttering the mantras of the Tattvas with the Mudra of morsels. Then the initiator acquaints him with the mantras of the Guru and Bala and the mantra of the candidate's choice or all mantras at once.

In doing the latter the initiator keeps his foot on the head of the candidate. He also gives him a new name of initiation consisting of two or three syllables with the word aanandanaatha coming at the end. Having thus brought him up to the same footing with himself in the spiritual sphere, the initiator should embrace him and smell him on the head. In order to keep his own spiritual power intact the teacher should also perform the Japa of the Mulamantra one thousand and eight times. The cadidate should offer due and costly presents so that his preceptor may be satisfied.

28th Chapter

In this are given the instructions regarding Purashcharana which is necessary for obtaining success in the mantra. Purushcharana means performance of the Japa of a mantra either one thousand and eight times or fifty thousand times. This is necessary to remove the obstacles due to lack of faith, scepticism and the demerits of the previous lives. This may be done either by the disciple or by the preceptor or by some worthy Brahmin in some sacred place either on the 14th or the 8th or the 15th day of the dark half of the month. The practitioner is to keep a fast for the previous day and perform worship to

Brahmins or teachers, Ganesha, mother deities and manes. The Japa is to be done in a suitable place which is to be selected carefully by testing the same with the help of the diagram called Kurmachakra consisting of nine sections in which the alphabet or the Matrika is arranged in a prescribed manner.

The rosary to be used is to consist of the beads of pearl, crystal, etc. Beads may number a hundred, fifty or twenty-eight. One of them which stands for the Meru should be the biggest, the rest being gradually smaller in size and each bead should have a knot. The holy articles of food, such as, curd, milk, Ghee, etc. should be taken and those of the opposed character should be avoided.

The practitioner should also observe sleeping on the earth, celibacy, silence, absence of envy, three daily baths, avoidance of mean acts, daily worship, daily charity, praising of the deity, occasional worship, reliance on the teacher and the deity and avoid the company of and eating with women, Sudra, outcast, etc., lying, cheating, deceitful talk, hearing the music and seeing a dance, etc. Japa should not be done when the doer is without a Pavitra in the hand, naked, with the head covered, without a seat, asleep, going or standing in the street, dark places or talking. It should not be continued on seeing, talking to and hearing an outcast, sneezing, discharging the wind and yawning. Everyday such an amount of Japa should be done from the morning till mid-day as has been fixed per day.

There should not be a pause of even one day in the period of Purashcharana. At the end of Japa the offering should be made to the extent of its 10th part in the purifying fire. Tarpana, Marjana and the feeding of the Brahmins should be done respectively to the extent of the 10th part of the Homa, 10th part of the Tarpana and 10th part of the Marjana.

The five acts of a sacred bath, purging of the sins, offering water to the Sun, sipping of water and Pranayama come after the above. The large scale worship is to be conducted at the end to complete the Parashcharan.

There is yet another way of performing the same. According to this the Sadhaka is to keep a fast previous to the day on which a Solar or Lunar eclipse fails and stand in the river navel-deep and mutter the mantra during the time of the eclipse, and after having done so perform Homa, etc.

29th Chapter

In the above the practitioner is instructed to mutter the mantra of Devi with his wife at the time of midnight, because anything done of religious merit at this time has got far greater significance than if done at others. Performance of Japa nine-hundred thousand times of Devi is calculated to qualify the practitioner for approach to her.

This is to be done in the following manner:

Three hundred thousand times when the practitioner has to concentrate his mind upon Kundalini in the Muladhara centre coiling herself round a phallic mass of light called Svayambhu. Again three hundred thousand times in the heart with the mind fixed on Kundalini with the Linga as above called Bana. Lastly the same number of times in the head with attention fixed on Kundalini with the Linga called Itara.

Mind loses its separate entity and becomes one with the object of meditation when Japa is performed a crore times. The serpent power lying dormant is to be awakened and raised from centre to centre in the body of the Sadhaka and these centres number nine and are located in the Muladhara, Linga, navel, heart, throat, middle of the eyebrows and so on. The practitioner is advised farther to see that the mautra when muttered is not subjected to the impurity, attendant on birth and death. Each breath before its beginning and termination is said to have this pollution. Immunity from such impurity is obtained by that practitioner who knows the meaning of the mantra, the life of the mantra and the Yonimudra.

The meaning of the mantra of Devi Tripurasundari is said to be sixfold, *i.e.* literal meaning, traditional meaning, inner meaning, Kaulika meaning, occult meaning and real meaning.

According to the literal meaning, the mantra is to be understood as implying the union of Shiva and Shakti because the constituent syllables of the same represent either Shiva or

Shakti. Hence the first part of the mantra called Vagbhava, *i.e.* ha sa ka la hriim represents Vama Shakti, Brahma, Goddess of wisdom, volition and Purvbamnaya, the second called Kamaraja, *i.e.* ha sa ka ha la hriim, Jyeshta Shakti, Vishnu, Goddess of earth, cognition and Dakshinamanya, and the third called Shakti, *i.e.* sa ka la hriim, Raudri Shakti, Rudra, action and Pachimnaya. The syllable at the end of all these parts represents Ambika Shanti, Shambhu, Santa, all the three aspects volition, etc. and Uttaramnaya.

All these parts are further viewed as representatives of the three acts of creation, sustenance and destruction. In this way in the Vagbhava with reference to the creation as beginning, ha denotes Brahma, sa Goddess of wisdom; with reference to the same as continuing (considered as sustenance) ka denotes Vishnu, la Goddess of earth; with reference again to the same as ending, ha denotes Rudrani and ra Rudra and the syllable ii denotes Shiva Shakti and their combination (Anakhya). In the Kamaraja ha and sa represent continuity as beginning, ka and la continuity as continuous, ha continuity as ending and ii as Anakhya. Addition of ha in this part indicates that it refers to the stage of continuity. In the Shakti, sa as abbreviation of ha sa denotes destruction as beginning, ka la destruction as continuity, ha destruction as ending and ii the same as before.

According to the Sampradayartha, ha denotes sky ka air ra fire sa water and la earth, giving birth respectively to sky, etc. The goddess of the mantra is to be meditated upon as identical with the whole universe. The fifteen syllables of the mantra respectively represent also the 15 qualities of the elements, *i.e.* one of the sky, two of the air, three of the fire, four of the water and five of the earth.

The mantra when analyzed into its constituents including both vowels and consonants resolves itself into the 37 parts as five vowels (four a + one ii) and six consonants ha sa ka la ha ra in the first part, as six vowels (five a + one ii) and seven consonants ha sa ka ha la ha ra in the second part, as four vowels (three a + one ii) and five consonants sa ka la ha ra in the third part, as three Bindus at the end of these parts and as a whole. These 37 constituents represent the 36 categories both individually and collectively. The six ha's in the mantra represent the sound as effect (fivefold) and as cause. The three ii's and the Bindu at their end stand for the four forms of touch, three ra's represent three aspects of form, two sa's represent two kinds of water, gross and subtle and the third. Sa stands for the connection between water and nectar. Three la's represent the earth as existing in the three worlds, three ka's also represent three classes of cognisers-Ashuddha, Shuddha and Mishr, 12 a's represent Jiva as such, Jiva as breath, Jiva as spirit, three Bindus, purusha, Raga, Vidya, Kala, Niyati, Kala, Maya, Shuddhavidya, Ishvara and Sadashiva and three Nadas, Shanti, Shakti and Shambhu.

According to the inner meaning, the mantra reminds of the identity between Shiva, the teacher and the disciple.

According to the Kaulika meaning it reminds of the identity between the mantra, the diagram, the deity, the teacher and the disciple. Identity between the mantra and the diagram is thus established:-

Three la's = three rectangles, three sa's = three circles enclosing the two lotuses, thrice three parts of hriim, *i.e.* ha ra ii = nine triangles, three ha's = one 14 sided figure and two decagons, three ka's = the centre.

Identity between the diagram and the deity is shown as below:-

One hundred and eleven deities that are worshipped in the different parts of the diagram attend upon the Devi and hence she is the main deity of the chakra. She is also manifest in the form of the planets counting nine as she constitutes the essence of the moon, the sun and the fire, volition cognition and action, Sattva, Rajas and Tamas.

She appears in the form of constellations representing sense organs (5) active organs (5) sense and action objects (10) mind, wisdom, ego, chitta, matter, qualities, and spirit.

The six Yoginis Dakini to Hakini as presiding deities of the six constituents of the body skin, etc. and the eight mothers Brahmi, etc. as presiding over vowels, gutturals, palatals, linguals, dentals, labials, semi-vowels and aspirates are mere representations of the Devi. As one with the ten vital currents, limited soul and supreme soul she appears as twelve signs of the zodiac.

The above forms of Ganesha are also traceable from the mantra. The three parts of the mantra respectively represent Vaikhari, Madhyama, and Pashyanti as presiding over the three groups of sixteen letters each (tha to sa, ka to ta, a to ah). Hence Devi the deity of the mantra stands as the mistress of all the forms of speech.

As the mantra has three parts, three Bindus and three Nadas, it reminds of her planetary form. Again as it possesses three parts having five letters (ha ra ii Bindu and Nada) at the end of each in addition to the remaining letters numbering 12 she is to be considered as one with constellations. As the six letters suggestive of Shakti, *i.e.* three hriim and three la's represent Yoginis she assumes their form. Similarly all the letters of the mantra excepting the three ending hriim counting 12 stand for the 12 signs of zodiac and hence she appears in the form of Rashis also.

The Identity between the Teacher and the Taught: This can be acquired and realized just as that between the Deity and her body either as a mantra or as chakra, which has been described above.

According to the occult meaning, the mantra reminds the practitioner again and again to be attentive to the idea of identity between his self and Kundalini which wakes up in the Muladhara and rises from there to the lunar seat in the head.

According to the real meaning the Sadhaka has to merge himself in the Transcendental Self.

30th Chapter

The three parts of the incantation of Devi known by the terms Vagbhava, Kamaraja and Shakti together with the Hamsa-mantra are described in this chapter. ha and sa of hamsa are said to be representative of the vital breath and the soul, existence and intelligence, Shiva and Shakti. This mantra is called Ajapa because it is not to be muttered but to be meditated upon so much so that the practitioner is to realize his unity with the Transcendental Self. Of the Matrika or the Sanskrit alphabet, vowels constitute the breath and Bindu and Visarga consciousness.

Every incantation becomes lifeless, as it were, if not possessing Bindu, Visarga, ha, sa and ii. shriim hriim hamsah aim kliim soham hriim shriim is supposed to vitalize every mantra as do Bodhini and Dipini. Meditation on Kundalini in the form of a flame, if accompanied by the three Kutas above referred to, leads the practitioner to acquire respectively extraordinary wisdom, power of charming and immunity from the effects of poison.

The meaning of Kamakala and what it stands for and other details regarding the same are given at the end of the chapter.

31st Chapter

This chapter gives the details regarding the Homa or sacrifice. These are offerings to be made, formation of the Kunda and the diagram, sacrificial pots, the incantations meant to invoke the fire, etc. Particular objects obtainable from making particular offerings are also fully mentioned together with the measure and weight of the latter at the end of the chapter.

32nd Chapter

How a particular ring is to be prepared and how the same is to be worn by the worshipper of Devi are mentioned in this chapter. The ring like the letters of the Sanskrit alphabet, is to be considered as identical with the Sun, Moon and Fire. Hence the same is to be made up of gold, silver and copper as representing Sun, Moon and Fire. The metals named above are to be in the ratio of 12, 16 and 10.

The ring is to be washed with the five products of cow and five Amritas on a bright Tuesday. The alphabet should be written in the form of a lotus and the nine pots full of perfume, water and jewels should occupy the place of the eight leaves and the centre of the lotus. In the central pot should be placed the ring the three parts whereof should be worshipped as Moon. Sun and Fire with the three Bijas of Devi and with the three groups of letters, 16 vowels, 24 consonants ka to bha and ten other consonants ya to ksha.

Nyasa of Sammohini and Matrika comes after this. The goddess of the latter is to be meditated upon as three-eyed, moon-white and wearing the emblems rosary, goad, arrows, noose, two lotuses, bow and book. After offering the worship with flowers, lamps, etc., together with that of attendant deities and muttering the mantra a thousand times, the ring should be put in hand. The Sampata Homa also should be done after this. The disciple who is to wear the ring should be bathed with the water of the nine pots above referred to when placed on the lotus of the Matrika. When thus qualified, he is to wear the same on the index finger. It may be made of nine jewels which are sacred to the nine planets which represent the nine classes into which the Sanskrit alphabet is divided.

33rd Chapter

This chapter deals with the invocation of the presence of Devi. It is said to be possible in three ways in the diagram, in the image and in the disciple. The disciple should make an image either of a jewel or of gold or of silver and have Devi invoked in it. The image should be taken out every month and worship conducted through it with the 64 articles of the same. The diagram should be washed with the five products of a cow placed in an altar on a conch. After this he should do either items of worship and having washed the same diagram with different juices and waters and put the same on a golden seat. Then Devi is to be invoked and the diagram retouched and electrified with the mantra aam hriim, etc. Then is the worship to be done with the necessary articles. Recitation of the hymns completes the worship.

34th Chapter

In this the reader finds the details given about the worship of Devi as done by the follower of Kulachara. Among many other things, he has to observe that he does not keep fasts, does not see a naked lady or one with visible breasts, that he avoids anything painful to women, that he eats modestly so as to keep healthy, that when sacrificing an animal he gets it beheaded by someone else while reciting the mantra udbudyasva, etc. He is further directed not to be too much given to hunting, nor to indulge in the gambling but to strictly watch the conduct of women at home.

He alone is to make use of the liquor who is not adversely affected by it but who on the other hand is uniform in treatment to all beings, is equally affected by honour or dishonour, by friend or foe, by stone or gold and who speaks smilingly and with steady eyes. The best sign of success in worship is afforded by detachment, thirst for release, renunciation, control over everything, the practising of the eight-limbed Yoga, avoidance of enjoyment, mercy towards all beings, acquisition of omniscience, etc.

35th Chapter

The worship of Devi through the medium of a lady forms the subject-matter of this chapter. This is strictly to be done in privacy and at night. In laying stress on the use of the necessary material for this sort of worship instances are quoted of Vishnu, Indra, Pushpadanta, Dattatreya and Balarama. Various powers are said to be obtainable from this worship. But the worshipper is to do every function with the help of his wife.

36th Chapter

In this chapter, the follower of Kulachara is instructed to have his wife initiated like himself by his own teacher who is to treat her as his own daughter. The Kula worship is to be celebrated on the 8th and 14th day of each half of the month through a diagram drawn on an earthen pot with red lead. The diagram is to consist of a hexagon inside a lotus. The image of Devi is to be drawn with the same material in the middle of the hexagon. The name of the person to be hypnotized is to occupy the cheeks of the image. The mantra of Devi aim kliim sauh comes before and after.

The Kamabija, *i.e.* kliim, should be prominent in the other parts of the image. The muttering of the Vasantasundari mantra kliim hriim aim niilasubhage hili hili vicche svaahaa comes after this. Then he has to electrify the image with the spiritual power and after having done the same perform the actual worship. The spiritual counterpart of the above is also detailed in this. The practitioner has to place himself in an even seat in a pure solitary place with the nervous system fully purified. He has to purify the latter with the Pranayamic exercises and be in the Sundari-mudra in which eyes are fixed on the nose-tip or the inner soul, or the middle of the eyebrows and the teeth, lips and tongue are to be closed with the serpent force drawn up from the Muladhara.

He has to melt the nectar in the thousand-petalled lotus in the head and with the nectar shower so obtained he is to worship Devi in the heart-lotus. Invocation of Devi is to be done through inhalation and retention of the breath. The same nectar is to be offered for the washing of the feet, Arghya and sipping. Thirtysix component principles of the world are to serve the purpose of perfume. The flowers to be offered are non-violence, abstention from theft, commiseration, forgetfulness, sense-control, presence of mind, absence of envy, hatred and attachment, and truthfulness. Air serves the purpose of incense, light of lamp, Sun of mirrer, Moon of umbrella, happiness of ornaments, sky of chamara and heart-throbbing of bell.

The nectar above referred to is the drink but not the liquor. By the Sundari or the matchless beauty is meant the subtle flame which occupies the central place of the Nada which ultimately is the essence of the fire, the air, the sky, the water and the earth. The practitioner of the worship has to merge the products of the five different elements in the elements themselves, lights in the light and the mind in the Brahman. In the pit of Kundalini he is to enkindle the spiritual fire and make into it the offering of all desires.

37th Chapter

This chapter relates how the things which have been rejected by the followers of the Vedas can be used by a Tantrik worshipper in a particular manner so as to gain success in worship. It is said that there is absolutely nothing pure and nothing impure. It is the way of using it that causes its purity or impurity. What is really needed is the fixing of mind on the Brahman. It is variously described as Prakriti by Samkhyas, as Avidya by the Vedantins, as Mahavidya by Shaivas and as Mahamaya by Tantriks. Brahma, Vishnu and Maheshvara adore Mahamaya. Brahma by worshipping her obtained Savitri representative of the will-power as his consort and Vishnu Lakshmi representing the cognitive power.

She is said to have two forms gross and subtle. The gross form is visible in females and the subtle in the Kundalini. It is therefore, commanded that the females should not be subjected to any cold treatment such as beating, etc. Worship of the subtle form is conducted through the Kundalini Yoga.

38th Chapter

In this Shiva says that the Devi feels satisfied not so much with the mental worship as with the physical. In accounting for the same it is said how she was once praised by Brahma and how she being pleased conferred on the Trinity, Brahma, Vishnu and Shiva, their respective powers—creation, sustenance and destruction and how she was born first as Sati and then as Parvati and how she advised Shiva to marry her as a Vira and how he praised her expressing that all his eight forms were really her own manifestations and at the end how she appeared before Shiva in the form of the Universe.

The reader is here to see for himself how much the Tantra has borrowed from the Durga-saptashati and the Bhagvad Gita.

39th Chapter

This chapter sets forth that spirit and matter are the two eternal principles. The former is represented by the male sex and the latter by the female. The latter, therefore, is endowed with the Gunas (qualities). From this view-point the question of purity and impurity attaching to the things of the world is said to be conventional for taking it for granted that the world consists of the five elements, we have to see how the same earth can be considered to give rise to the cast distinction such as, Brahmin etc., or to the distinction either as divine or as human, etc. He is held to be a Brahmin who feels really that the soul is different from the body and is one with the Brahman as Supreme Bliss. The Vedantic description of the Brahman and of the bodies of the soul, gross, subtle and causal is put in the mouth of the Devi.

40th Chapter

In this Shiva and Parvati change positions. Shiva becomes the questioner and Parvati the replier. The latter states how the world consists of the two elements Shiva and Shakti and how their simultaneous existence is essential and how she is to be worshipped through incantation, image, etc. When physical worship is to be conducted the Linga is advised to be worshipped alongside. The way of worship associated with the Kulachara is fully detailed towards the end.

41st Chapter

In the above explanation of the Anuttara or the Transcendental Self comes first. It is defined as the unification between Sat and Chit, Prakasha and Vimarsha, Shiva and Shakti (being and knowing). The Bija of the Devi given here points to the same thing so far as its connotation is concerned. All the incantations are said to lead when practised to the same stage in the long run, though at first they may be helpful to the practitioner's approach to Shiva or the Devi. The practitioner has to bear in mind that the objective world is in nature one with the subjective. The Transcendental Self manifests itself trebly, first as Bindu, then as Nada and lastly as Bija. The Shaivistic doctrine regarding the cosmic evolution and that of the Sanskrit alphabet is described in elucidation of the Anuttara much in the same way as Abhinava Gupta and Kshemaraja have done.

The verse frequently quoted by the writers on Shaivism comes at the end and means an individual soul out on gaining oneness with the supreme reality as supreme bliss energises to merge the objectivity in the subjectivity.

42nd Chapter

This closes the book and deals with the description of the physiological and spiritual phenomena as known to the experts in the Tantrika lore. It is given in the form of question and replies. Devi asks the following questions: 1. Whence does the speech originate and wherein does it disappear? 2. Who feels appetite and thirst and who sleeps and wakes up? 3. Who sins and is bound and who emancipated? 4. How is the etheric side of the soul to be accounted for in the body? 5. How does the soul assume the physical body? 6. Who is the soul? 7. How does the soul see? 8. How does it become Sakala (bound) and how Nishkala (released)? 9. How does it get sustenance and what accounts for its physical appearance and disappearance?

Shiva replies: 1. Speech comes from and merges in the mind through the instrumentality of Avyakkta and the vital breath. 2. Vital breath feels desire for food and drink and bodily fire enjoys them. Wind (Vital breath) causes sleep and waking. 3. Mind is responsible for sins and bondage. When the soul being under control ceases to function, it becomes free. 4. The soul has three sides, aerial, astral and etheric as breathing, digesting and speaking. 5. The soul appears in the physical form under the influence of matter as Sattva, Rajas and Tamas and it is centred in the tip of the nose and heart.

It lives as embodied so long as the heart and the head continue to function. 6. The supreme reality is the soul. 7. It sees through the body. 8. It becomes Sakala when it comes under the influence of Prakriti and Nishkala when it realizes its unity with the Trancendental Self. 9. It gets sustenance through its actions and appears and disappears through the presence or otherwise of limitations caused by action. It rises and falls like a ball in the scale of life.

In explaining the above replies Shiva makes a reference to the doctrine of the Upanishads as revealed to Sanatkumara the son of Vishnu. The last lines of the chapter like those of the Upanishads, describe the merit of reciting the Tantra.

Jnanarnava Tantra

This work falls into the Shri Vidya class of tantras and is a relatively brief and comparatively straightforward example of the genre. Jnanarnava means Ocean of Knowledge.

Consisting of 26 patalas (chapter), the Jnanarnava amplifies information relating to the Shri Vidya tradition in other works of the school. No date can be assigned to it. It does have some interesting information on inner worship, rather than the external rituals (bahiryaga).

Chapter One

Cast in the familiar agamic form as a discourse between Devi and Ishvara, the goddess starts the Tantra by asking of the true nature of Shri Vidya. Ishvara opens by saying that the letters of the Sanskrit alphabet from A to Ksha, endued with the 14 vowels and the three bodies, constitute the body of Matrika Devi, which is one with the circle of time (Kalamandala). This is the absolute as sound (Shabdabrahma). It is the true form of the Atma and is Hamsa. Ishvara then dilates on the three bindus. Hamsa indicates the three gunas; the three Shaktis Iccha, Jnana and Kriya; the three tattvas; the three cities; the true nature of Bhur-Bhuvah-Svah and the states of waking, dreaming and deep sleep. The waking state is sattvik, and is the true form of Shakti, while the deep sleep state is tamasik and is the Shiva form. The dream state is rajasik, says Ishvara. The Turiya (the fourth) pervades all these states and is the supreme Kala, the Jnanachitkala, a state of true consciousness. This is Tripura and the true rosary of the letters (akshamala).

Chapter Two

Devi wants to know about the different mantras and forms of Tripura. Ishvara says that Tripura has three forms of which the first is Bala. He outlines a three syllable mantra which is Aim Klim Sauh. She is the mother of great good fortune, the giver of great eloquence, the great destroyer of death and consists of all the worlds. The rest of this chapter is mainly given over to details of external puja (worship), including tantrik gayatri and the other usual details including nyasa and the like.

Chapter Three

Ishvara starts to talk about internal worship (antaryaga). Before beginning external worship, the devotee should visualise the root mantra as pervading the body from the Muladhara to the top of the head, and visualise it as effulgent as kotis of fires, suns and moons. Then, facing east, the devotee should inscribe the yantra.

The text gives detailed instructions on how to draw the figure, including the mantras and bijas which should be drawn on it. The yantra may be inscribed on gold, silver, copper or on the ground, and should be scented with perfumes including sandalwood, as well as coloured with kumkum, vermilion and camphor. It may also be inscribed on bhurja (a type of birch bark). Then follows a detailed dhyana (meditation image). She is adorned with many strings of pearls, a bright diadem. In her two left hands she holds a book and a bow, and with her two right hands she banishes fear and bestows boons to the sadhaka. She is pure white as milk or snow and has a sweetly smiling face.

Chapter Four

Here, Devi asks Ishvara about how to perform puja in the chakra or mandala described earlier. Shiva describes twelve pitha shaktis whose names are Vama, Jyeshtha, Raudri, Ambika, Iccha, Jnana, Kriya, Kubjika, Riddhi, Vishaghnika, Dutari and Ananda. They are adorned with strings of pearls and rubies, resemble the moon, are as white as the Ganges river in flood, and have two arms. The twelve shaktis are to be worshipped from the east (of the goddess) and are adorned with the nine jewels.

There follows a description of the five corpses, the five lion seats. Ishvara says that these are the bodies of Brahma, Vishnu, Rudra, Ishvara and Sadashiva. Brahma, Vishnu and Rudra represent the three gunas and the states of creation, maintenance and destruction. There then follows a detailed description of different mudras and other ritualistic details in the puja, as well as a description of the other deities who receive offering in the yantra.

Chapter Five

The subject of puja is continued here. There is a very lengthy meditation image of the goddess, followed by passages about the worship and the mantras of the yoginis, Batuka, and the other bali devatas, as outlined in Gandharva Tantra. The five great corpses form the base of the Hamsa mattress.

Chapters Six to Nine

Chapter six deals with the eastern lion seat, and describes the different Devis and shaktis who dwell in the direction. The mantra of Tripura Bhairavi is given. The Tripura Bhairava vidya is said to be hard to obtain in the three worlds.

Sampatprada Bhairavi is the great giver of prosperity. She is as bright as a thousand suns, with a crest gem like the rising moon, wearing numberless gems and pearls. Her face is like the full moon, and she has three eyes, with large swelling breasts, wearing red clothes, and has a youthful, intoxicated form. She holds a book and dispels fear with her left hands, while with her right she holds a rosary of rudrakshas and shows the mudra giving boons. There then follows a description of Chaitanya Bhairavi. The next chapters describe the southern, the western and the northern directions. Although the dhyanas differ from those given in the Paradise Island section, the different devis have similar forms. Here their yantras and vidyas are also outlined.

Chapter Ten

This is a lengthy chapter which covers a number of mantras necessary to the worship of the Devi. These include the hand-purifying mantra (karashuddhi), the asana or seat mantras, and the other vidya mantras used in her worship.

Chapter Eleven

A very brief chapter of only 14 verses which describes the Panchadashi (fifteen letter) Kadi vidya, all in code form.

Chapter Twelve

Describes the other divisions in Shri Vidya, starting with the Lopamudra mantra.

Chapter Thirteen

Deals with the Sodashi Devi, whose mantra has sixteen letters. It describes the different sections of the mantra and says that the sixteenth letter should never be revealed to anyone. Unless it is obtained from a guru, its use bestows a curse. It is made up of four parts which correspond to the states of waking, dreaming, deep sleep and the fourth state. The fourth state, Turiya, is the supreme kala, above being and non-being, above the gunas and pure.

Chapter Fourteen

Describes the placing of the golden vessel or jar, and details the ten kalas of Fire, the twelve kalas of Sun, and in the moon mandala, the sixteen kalas, together with the mantras used in this worship.

This patala speaks then of the special offering, and of the yantra for this offering, which is made up of a triangle, a circle, six angles, and an earthsquare or bhupura. The sun is to be worshipped in this yantra.

The different six limbs of nyasa are also worshipped in this yantra, and a sadhaka should perform bhutashuddhi, or purification of the elements in the human body.

Different more complex nyasas should be performed, including the Mahashoda Nyasa, which is elsewhere on this chapter, and detailed in the Yoginihridaya. The different meditation images to be used are outlined. The puja includes the visualisation and placing of the 50 (51) letters of the Sanskrit alphabet.

This is a very lengthy chapter, much of which repeats information about puja which is rather neatly summarised in the Subhagyodaya.

Chapter Fifteen

Ishvara starts by saying to the Devi that he will declare the utmost Nitya mandala, and starts by speaking of Kameshvari, the mahavidya who subdues all of the worlds. Her mantra is then given in code form.

The vidyas of the other fourteen Nitya devis are then related. Mudras and nyasas related to Lalita are outlined. Inner worship (antaryaga) is recommended over external puja.

Chapter Sixteen

The text then proceeds with a question from Shri Devi about inner yaga and outer yaga (worship). Devi, replies Shiva, exists between the muladhara and brahmarandhra chakras.

There then follows a beautiful and lengthy dhyana (meditation image) of the queen of queens, who has a face like the full moon, a mouth like a lotus, and who consists of all mantra, all agamas, all places, all vidyas, all worship and puja, all shastras or holy texts, all amnayas, and who is pure bliss and consciousness herself, the supreme Mother. She should be invoked in the centre of the chakra using mudras, and all the mudras should be shown to her. She is surrounded by her fifteen Nitya attendants, and by the nine gurus, all of whose names end in Anandanath, as well as by masses of enlightened gurus. She is the Navachakreshvari, or Lady of the Nine Chakras. Then follows a very lengthy description of her other attendants in the Shri Yantra, much of which material is outlined elsewhere on this chapter, as well as her nine forms in the nine mandalas of the Shri Yantra.

Chapter Seventeen

Shiva is asked about the sixteen letter vidya, about the rules of reciting the mantra, as well as some instructions for the substance the Shri Yantra may be drawn on, which include bhurja bark, gold, silver, copper and the like. This chapter also has some prayogas (applications) for subjugation and the like, and the flowers and other substances used to obtain the desired results.

Chapter Eighteen

This short chapter deals with a rite known as the ratna or jewel puja, which is also described in the Gandharva Tantra. Performing this rite for a period of one month removes the blemishes accrued from seven incarnations, the text claims.

Chapter Nineteen

Shakti asks the lord to explain the nature of the three bijas, the essence of Tripurasundari. There follows a dhyana of Kamakala, which, the text says, deludes the entire world, and delivers every other type of benefit, including destroying death, and so forth.

Chapter Twenty

Deals with the rules of japa and homa, and describes successive homas which involve the recitation of mantra many hundreds of thousands of time.

This chapter is reminiscent of a similar chapter in Vamakeshvarimatam. The successive recitations and the more intensive homas eventually cause every denizen of the three worlds to become attracted to the sadhaka. It also describes the construction of the kundas (fire pits) to be used in these homas, together with other diagrams employed in the homas, along with the substances such as camphor and kumkuma which should be used in the worship.

Chapter Twenty One

Speaks of inner (antar) homa, which is figuratively described as using a four square kunda.

In this inner homa, the 21,600 breaths of inhalation and exhalation feed the fire of consciousness. These breaths make up the embodied being, or jiva, which, however, is one with the atman.

By sacrificing everything in the microcosm, which is one with the macrocosm, into the central sushumna fire, knowledge (Jnana) is obtained.

Chapter Twenty Two

Feeding and worship of the kumaris or virgins is the topic of this chapter. The kumaris should be treated with great reverence and fed good food, adorned with jewels and fine clothes and the like. Following the rules relating to the kumaris, the chapter then goes on to describe the Duti, or shakti of a sadhaka. This section of the chapter contains reference to the virasadhana.

Chapter Twenty Three

Continues the topic of Dutis (messengers) by speaking of the Inner Duti. She is Iccha, Jnana and Kriya Shakti, the self of Shiva and Shakti, the parabrahma, or supreme absolute, in which everything is dissolved. She is the form of the sixteen vowels of the alphabet, the form of the absolute, and the sixteen kalas. Whosoever knows her through the grace of the guru, becomes one with her. A fine verse towards the end of this chapter says that there is no difference between the four varnas (so-called castes) and the chandala. The Duti, or inner goddess, is free from such distinctions. In her, all such distinctions do not exist.

Chapter Twenty Four

Ishvara starts to speak about the rules for initiation (diksha), without which worship of the Devi is fruitless. Would-be candidates who do not have the right attitude are not suitable for initiation. He then speaks of a pavilion to perform the initiation, and of diagrams to be created and of the devis who rule over the initiation, as well as the regulation of breath (pranayama) and other particulars of the time of initiation.

The chapter contains an unusual description of the chakras and of initiation in these chakras, as well as giving the right times to bestow initiation upon a candidate, which refer to particular times according to the rules of sidereal astrology when initiation will be successful.

Chapter Twenty Five

A very brief chapter which contains the rules for making a pavitra. This sacred thread must be scented with rochana, kumkuma and the like, and fashioned in particular way with 118 threads intertwined together.

Chapter Twenty Six

The last chapter in the Tantra speaks of the damana rite, which must be preceded by elaborate precautions to protect the sadhaka performing it.

Lalita Tripurasundari, the Red Goddess

What is Shri Vidya and what relationship does it have to the goddess Lalita and to her yantra, the Shri Yantra? Vidya means knowledge, specifically female knowledge, or the goddess, and in this context relates to her aspect called Shri, Lalita or Tripurasundari whose magical diagram is called the Shri Yantra. She is a red flower, so her diagram is a flower too.

The tantrik tradition views its symbols as having a gross aspect, a subtle aspect, and a supreme aspect. In terms of Lalita, the gross form is the image of the goddess with her four arms and so forth, the subtle form is as yantra, and the supreme form is her mantra, all three being the goddess in different aspects. Behind the sometimes colourful symbolism is deep wisdom coupled with pratised methods for realising oneself.

Lalita loves puja. This term is usually translated as worship. However, this is misleading, as it introduces a duality into a process intended to bring the practitioner (sadhaka or sadhvika) to a non-dual position. There can be various pujas including daily rites, those performed at the four twilights, rites done for specific objects, optional rites done on festival days, or on otherwise auspicious days such as lunar eclipses or the entrance of the sun into a sidereal constellation, rites in assemblies or groups, and rites accomplished with a partner. Subhagodaya, on this chapter, is a translation which gives the full puja of Tripurasundari or Lalita.

Lalita means She Who Plays. All creation, manifestation and dissolution is considered to be a play of Devi or the goddess. Mahatripurasundari is her name as transcendent beauty of the three cities, a description of the goddess as conqueror of the three cities of the demons, or as the triple city (Tripura), but really a metaphor for a human being.

What then is a yantra? The word is usually translated as a machine, but in the special sense of the tantrik tradition refers to the Devi in her linear or geometrical form. Yantras, by the way, are always used flat. They may be two-dimensional or three-dimensional. Every aspect of Devi has her own mantra and yantra. The yantra of Devi Lalita is Shri Yantra. The divinity of the yantra always occupies the centre or apex.

The different parts or petals and lines of the yantra are usually arranged in concentric circles (mandalas) and contain rays or sub-limbs of devi. The Shri Yantra has nine of these mandals, each filled with various aspects of the Devi. In Shri Yantra there are 111 aspects. The Shri Yantra is said to be a geometric form of the human body, which implies that goddess as Macrocosm is one with human being as Microcosm.

Formation of the Shri Yantra: The creation of the Shri Yantra is described in the Yogini Hridaya (Heart of the Yogini Tantra), which still does not exist in an English translation, as far as we are aware. This is said to be the second part of the Vamakeshvara Tantra.

From the fivefold Shakti comes creation and from the fourfold Fire dissolution. The sexual union of five Shaktis and four Fires causes the chakra to evolve. O Sinless One! I speak to you of the origin of the chakra. When she, the ultimate Shakti, of her own will (svecchaya) assumed the form of the universe, then the creation of the chakra revealed itself as a pulsating essence. From the void-like vowels with the visarga (:) emerged the bindu, quivering and fully conscious. From this pulsating stream of supreme light emanated the ocean of the cosmos, the very self of the three mothers.

The baindava of the chakra has a triple form, dharma, adharma and atma, and matri, meya and prama. The chakra of nine yonis is the great mass of consciousness bliss and is the ninefold chakra and the nine divisions of the mantra.

"The baindava is placed on a dense flowery mass and is the Chitkala. Similarly, the Ambika form of eight lines is the circle of the vowels. The nine triangles quiver forth the effulgent form of 10 lines. The Shakti, together with her surrounding nine blossomed forth the 10 trikonas. The second quivering form of 10 lines has Krodhisha as first of the 10. These four chakras, of the nature of light, create the 14-fold form, the essence of perception."—Yogini Hridaya, I 6-16.

At the very heart of the bindu or centre of the Shri Yantra is that which caused it to emanate. This is Kamakala, consisting of the three bindus or potentials. One is red, one is white, and one is mixed. The red bindu is ova, the white bindu semen, and the mixed bindu the union of Shiva-Shakti, the individual as potential Shri Cakra.

Father and Mother are represented in Shri Vidya by two limbs or aspects of Lalita known as Varahi and Kurukulla. The semen of Varahi, the father-form, gives four alchemical dhatus to the child. The ova of Kurukulla, the mother-form, gives five dhatns to the child. Consciousness enters via orgasm. The three bindus, collectively known as Kamakala (digit of sexual desire), are the root potential of sun, moon and fire. It is like sun and moon coming together in an eclipse, or the seed from which the plant human being grows.

Varahi's four alchemical dhatus are known as the four fires. Kurukulla's alchemical dhatus are known as the five saktis. The combination of these five saktis (downward pointing triangles) and four fires (upward pointing triangles), forms the complex figure in the centre of Shri Cakra.

Varahi's four fires are the 12 (3 x 4) sun Kalas, 12 sidereal constellations. Kurukulla's five triangles are the 15 (5 x 3) Kalas of the moon, 15 lunar days. The complete individual grows within nine months to be born as a Shri Yantra or plant. The flowering of this plant is shown by the 24 petals of the yantra. The above all gives rise to the familiar shape of the Shri Yantra.

The yantra is usually arranged in one of two forms. In the Bhuprastara, it is two dimensional and laid flat, usually facing the east, but sometimes the north, depending on the practice. The Meruprastara has the yantra in a pyramidal form. Unless the yantra be decorated with the appropriate bija and other mantras, it is worthless. It is also dead unless it is installed with life and the individual doing the puja is initiated into one of the lines (parampara).

Shri Kurukulla Devi

Devi Kurukulla is, like Varahi, a balidevata, that is a receiver of offerings. She is one with Lalita and is identified, at least in the Tantrarajatantra, with Tara. She is one with the 15 Nitya devis of the waxing moon. She has three mantras, one of seven syllables, one of 13 and one of 25 syllables but in her daily puja the mantra of Tara is employed (Om tare tuttare ture svaha).

According to the vast mediaeval tantrik digest Shrividyarnava, these mantras are Om kurukulle svaha (7), kurukullayah Om kurukulle hrih svaha (13); kurukullayah Om kurukullehrih: mama sarvjanam vashamanaya hrim svaha. Because she has three mantras she is called Trikhanda (three sections).

Her meditation image is as a naked woman, with dishevelled hair, red and full of bliss. Her four hands bear arrow, bow, noose and quiver. Around her is a multitude of Shaktis who look just like her and all have taken their first bath after menstruation, have flowering yonis and eyes intoxicated with desire.

The preparatory act (purashcharana) for worshipping Kurukulla is to recite her vidya-mantra 100,000 times, using red flowers.

Her yantra (top left) is inscribed with the 25 syllable mantra-vidya for success.

According to the tradition, yantras which do not have bija mantras inscribed on them, are powerless.

The Nine Mandalas of the Shri Yantra

The Earth Square or Bhupura: This mandala represents the enclosing walls or fence of the zonule of a practitioner. The three lines of the bhupura of Shri Yantra each has a set of subsidiary aspects or sub-limbs of the goddess. On the outer line are the eight world protectors (lokapalas), the guardian spirits of the directions and intermediate directions.

On the middle line are eight Siddhi Saktis identified with the senses. On the inner line are eight Shaktis ruling Desire, Anger, Envy, Delusion, Greed, Jealousy, Virtue and Vice. They are the eight Matrikas. These saktis are collectively known as the Obvious Ones (Prakata Yoginis). A form of the triple Devi known as Tripura rules all these shaktis in this individual mandala of the yantra known as 'The Chakra Ruling the Three Worlds'. She has four arms, is the colour of crystal, is adorned with pearls and holds a book, a pot, and a beautiful lotus, her Vidya is Am Am Sauh.

Outer Line: Indra (E) wears yellow, rides an elephant; Agni (SE) wears red, rides a ram; Yam (S) wears black, carries a staff; Nirriti (SW) wears dark green; Varuna (W) wears blue, and his vehicle is a makar; Vayu (NE) wears pale clothes; Soma (N) wears pure white; Ishana (NE) is a form of Mahadeva Shiva.

Middle Line: The Siddhi Shaktis are smeared with vermilion, wear red garlands, carry noose and goad, and are as bright and beautiful as red lotuses.

Inner Line: Brahmi wears yellow, has four arms, is beautiful. One hand dispels fear, one grants boons, the others hold a jewelled jar and makes the gesture of purification. Mahesvari wears white, has three eyes, holds trident, skull, axe, and vessel containing sour curds. Kaumari wears yellow, holds shakti-dart, Javelin, and makes the gestures of dispelling fear and granting boons. Varahi is dark in colour, holds conch, discus, dispels fear, grants boons. She wears many ornaments and gems. She has the head of a pig, holding plough, mace, sword and shield. Indrani is black, carrying a bright blue lotus. Camunda is black, holds trident and damaru (hourglass drum), holds axe, and milk in a bowl. Mahalaksmi wears yellow, holds serpent, shield, bell and milk in a skull shaped cup.

The nature of this outermost mandala is fire of fire. The gem is topaz. The time is 24 minutes (360 breaths). The Mudra is the All Agitating.

Sixteen Petals: The saktis in this circle are known as the Hidden Ones.

The whole mandala of 16 petals is called 'Fulfiller of Desire'. The presiding form of the Lalita is Tripureshi. Her vidya is Aim Klim Sauh. She is described as ornamented with all gems, carrying a book and a rosary. The 16 yoginis in this mandala are associated with the attainment of desires by the cultivation or strengthening of power over mind, ego, sound, touch, sight, taste, smell, intellect, steadiness, memory, name, growth, etheric body, revivification, and physical body. They are described as the Nitya Kalas. Each holds a noose, a goad, pot full of nectar, and makes the sign of giving. They are very red.

The gem of the mandala is sapphire. The dhatu of physical alchemy is chyle, the first product of the disintegration of food by the biological fires. The time is three hours (2700 breaths). The Mudra is the Wettening Mudra. The nature of the mandala is sun of fire.

Eight Petals: The saktis in this mandala are called the Very Secret Yoginis. The whole circle of eight petals is called the 'All Exciting Cakra'. Presiding here is Tripura Sundari. Her vidya is Hrim Klim Sauh. She is described as swaying because of her love intoxicated state, with her eyes full of bliss.

She smiles with passion. She shows the mudras dispelling fears and granting boons.

The eight saktis in each of the eight petals of the mandala are described as saktis of Speech, Holding, Walking, Excreting, Pleasure, Abandoning, Concentration and Detachment. They are described as sapphire blue, holding noose, goad, dispelling fear, and holding blue lotus. Their names (Ananga Madana etc.) all convey terms of loving sexuality.

The gem is cat's eye. The dhatu is Flesh. The time is day and night (21600 breaths). The mandala's nature is moon of fire.

Fourteen Triangles: This mandala is called 'The Cakra Bestowing All Good Fortune'. The Yoginis are called 'Concealed by Tradition'. The presiding form of the devi is Tripura Vasini. Her vidya is Haim Hklim Hsauh.

She is very red and very beautiful. Fourteen shaktis of the triangles are associated with the chief nadis or currents of bioenergy. They are described as being proud, wanton, young, colour of cochineal, ornamented with gems, holding noose, goad, mirror, winecup full of nectar. They are the Akarshanis or Attractors.

The gem is coral. The dhatu is blood. The time is weekday. The Mudra is called All Subjugating. The nature of the mandala is fire of sun.

Outer 10 Triangles: This mandala is called 'The Cakra Bestowing All Objects to the Sadhaka'. The saktis are called the Kula Kaulas. The presiding aspect of Red Devi is Tripura Shri.

Here, the goddess is as effulgent as 1000 rising Suns, adorned with celestial ornaments, with large rising breasts, holding book and rosary, dispelling fears and granting boons.

The 10 shaktis in the triangles are described as having thrilled faces, holding noose and goad and adorned with various crystal and heavenly gems.

These are the Yoginis of the 10 vital breaths. The gem is pearl. The dhatu is Ova/Semen. The time is Lunar Day (tithi).

The Mudra is called the All Intoxicating with Love. The nature is sun of sun.

Inner 10 Triangles: The mandala is called 'The Cakra Protecting All'. The Yoginis are called Without Origin. The presiding aspect of Lalita is Tripura Malini. Her vidya is Hrim Klim Blem.

She holds noose and goad, dispels fear, and holds a skull. She is of vermilion brightness.

Her shaktis are the colour of 1000 rising suns, adorned with pearls and gems, holding noose, chisel, and showing the gestures of knowledge, and giving boons. They are the saktis of the 10 Vital Fires. The gem is emerald. The dhatu is Marrow. The time is Lunar Fortnight. The Mudra is the Great Goad. The nature is moon of sun.

Eight Triangles: This mandala is called 'The Cakra Destroying all Disease'. The yoginis are known as the Secret or Rahasya yoginis. The presiding aspect of the Red Goddess is Tripura Siddha.

Her vidya mantra is Hrim Shrim Sauh. She is described as the Destroyer of Poison.

Her yoginis are the colour of pomegranate flowers, wearing red clothes, smeared with red scent, each carrying five arrows and a bow. These saktis are the rulers of Cold, Heat, Happiness, Sorrow, Desire, and the three gunas Sattvas, Rajas, Tamas. They are also called the eight Vasinis and rule the eight Sanskrit letter groups. The gem in this mandala is diamond (Vajra). The time is month. The Mudra is Khecari Mudra. The nature of the mandala is said to be fire of moon.

The Four Weapons: In between the mandalas of eight triangles and the central triangles are the four weapons of the Red Goddess—flowery bow, flowery arrows, noose and goad.

Central Triangle: This mandala is called 'The Cakra Giving All Success'. The Yoginis are called Very Secret. Lalita dwells here as Tripura Amba, her Vidya being Hsraim Hsrklim Hsrsauh.

She is also known as Sampatprada Bhairavi, coppery effulgent, like 1000 suns, with three eyes, a face like the moon, adorned with white gems, with a beautiful figure, rising swelling breasts, intoxicated, wanton, young, proud, holding book, dispelling fear, holding a rosary and granting boons.

Her three saktis are called Lady of Lust (Kameshvari), Adamantine Lady (Vajreshi), and Flowery Vagina (Bhagamalini). Kameshvari is called the Rudra Shakti. She is white in colour, besmeared with camphor, adorned with pearls and crystal, and various other gems, holding book, rosary, bestowing boons and dispelling fear.

Vajreshi is the Vishnu Shakti. She is bright as red powder (kumkuma), adorned with flowers and gems, like the dawn sun. Her eyelids are smeared with sapphire dust, she holds sugarcane how, flowery arrows, bestows boons, dispels fear.

Bhagamalini is the Brahma Shakti. She is effulgent as molten gold, adorned with priceless gems, holds noose, goad, and shows the gestures of knowledge and bestowing boons.

The gem of the mandala is Gomaya. The dhatu is Fat. The time is season (two months). The Mudra is the Bija Mudra. The nature of the mandala is sun of moon.

Bindu: This mandala is called 'Purely Blissful'. The Yogini in this mandala is the Queen of Queens, Rajarajeshvari, the Very Red One, her Transcendent Majesty Lalita Maheshvari Mahatripurasundari.

Her vidya (Kamaraja vidya) is ka e i la hrim ha sa ka ha la hrim sa ka la hrim, plus a secret 16th syllable. Her description is that given in Vamakeshvara Tantra. Surrounding her are the Fifteen Nityas. The gem is ruby. The dhatu is hair. The time is year. The mudra is Yoni Mudra. The nature of this central mandala is moon of moon.

Yantra Mantra Tantra of Lalita: Lalita, as primordial devi, rays out her attendants and shaktis as modifications of moon, sun and fire. In this Shiva has no place, no qualities, is without the ability to act. Only when united with devi may 'he' act.

This is based on the subtle and pratised idea of Shiva as pure consciousness, witness of the triple manifestation of his Shakti. This Shakti, the very essence of the three gunas of Sattvas, Rajas, and Tamas, is the cause of all manifestation in the universe and as a human being. The three shaktis, by blending and reblending, create all things.

Shakti is triple as sun, moon and fire—that is to say of all the sidereal constellations and planets, and therefore of Time itself. She is triple as Will (Iccha), Knowledge (Jnana) and Action (Kriya). She is threefold as intellect, feelings, physical sensation.

Shakti is triple as wake-dream-deep sleep. What is called the Fourth is the witness, Shiva, who is said to pervade the whole cosmos just as heat pervades a red hot iron.

The physical body, according to the precepts of Ayurveda, is triple as the 'humours' Vata, Pitta and Sleshma. The varying combinations of these three shaktis make up the physical body.

Shakti is also fivefold as aether, air, fire, water and earth. The combination of the five elements and three gunas produce Lalita's Eternities (Nityas)—15 in number, each identified with a lunar day of the bright fortnight. The moon, symbolising Shakti, is the mirror or reflection holding together all creation.

A close examination of the details relating to the nine mandalas of Shri Yantra reveals that the shaktis of the whole circle represent the human being, who, in potential, is Shakti-Shiva united. The aim is for a person to realise that all powers, energies and manifestation are shaktis of consciousness, pure awareness.

The yantra may be examined in two ways, either as manifestation or dissolution. Maintenance is an intermediate state between the two polarities. When she is worshipped as creatrix the order is from centre to perimeter. As dissolver, the puja is from perimeter to centre.

In Shivananda Yogi's Subhagodaya is given the daily ritual or puja of Lalita's Shri Yantra—based on the Vamakeshvara Tantra. This rite is based on non-dualism, in a spiritual sense the realisation of the intrinsic oneness of macrocosm and microcosm.

As the puja is intended to banish all thoughts of difference, the devi is first felt or visualised in the heart, and then drawn out via the breath and installed in the yantra. She is then worshipped as actually residing there. But a clear link has been made between subject and object. The true home of devi is as cosmo-creatrix in the heart of the body which is the devi in human form.

The Various Mandalas of Shri Yantra: The Triple Goddess, from her own will to manifest, extends herself in a ninefold way, as modifications of moon, sun and fire. The attributions of the various mandalas shows the type of energy represented. The meditation in Bhavana Upanshad is a figurative way of describing this celestial city or mountain which is a human being.

The island of jewels is the gross human body with its 9 alchemical bases or dhatus. Each is figuratively described as a gem—diamond, emerald, sapphire, ruby &c. The sea of nectar (semen/ova) is the base for the arising of the human body. The diagram suns up the meditation. We can see that this island of gems is a very pleasant place to he, full of gardens, with a beautiful, begemmed palace, wafted with a gentle breeze upon which is carried great fragrance, cool, alluring.

This indicates the Kaula view that one gains liberation by a very pleasant way, enjoying as one goes. This paradise island is very, very close. Each of the elements in the island meditation has a subtle meaning associated with the esoteric physiology of Shri Vidya.

She, Lalita, united with Shiva, is subtlety of subtlety, hidden behind the curtain hanging from the canopy. Her forms may appear to become progressively less subtle, but she still remains herself.

Bala-Sundari-Bhairavi: Although Tripurasundari, as mother of the universe (jagadamba) is the aspect most often met with in works of Shri Vidya, she is also worshipped as Bala (a young girl), and as Bhairavi (a crone).

As Bala, she is 16 years old, a virgin, very playful and dear. Bala has her own yantra and mantra, her vidya is Aim Klim Sauh.

Bhairavi is also an aspect of Lalita, but represents Shakti in whom menstruation has ceased, and has some affiliations with Kali.

Applications of Shri Vidya: There are many prayogas (ritual uses) related to Shri Yantra. Some rites depend on auspicious times, such as Full moon days or nights in specific solar months

Devi also manifests as the five elements of aether, fire, air, water and earth. The saktis are purple (air), white (water), red (fire), yellow (earth), blue (aether).

Chapter II of Vamakeshvara gives a large number of rites, which one is not entitled to perform unless the daily rite is also accomplished. These rites are called the shatkarma, six acts: protection, peace, victory, wealth, punishment, destruction. The categories vary occasionally. When punishing an enemy it is necessary to both protect yourself and to know the right time for performance, according to the rules, and also the vulnerable points, which vary with the phase of the moon and with astrological aspects.

It is important to remember that Shri Vidya was primarily oral, and vital information was often left out of the written versions, so it is necessary to know a host of things before a rite can be started.

Initiation: Devi Lalita may be installed in a disciple, a yantra, or an image. All the methods essentially follow a similar form, but the right time must be selected. A disciple must have the necessary qualifications and potential. After initiation, she or he is to perform an operation to endue the vidya with energy or life. This involves the recitation of the root vidya a specified large number of times, although other valid methods exist for preparation.

The Philosophy of Tripura Tantra

It is a truism that every system of theocratic culture in India has behind it a consistently evolved system of philosophic thought. It is difficult, in the present state of our knowledge, to give a definite idea of the number of such systems in ancient and mediaeval times and even of the extent of literature comprised under each. Continued progress in researches in this field is likely to yield fresh materials favourable to the better understanding of the true history and philosophic value of these systems. The work known under the name of "Tripurarahasya" (Jnana Khanda) forms indeed a highly important document in the history of Indian Philosophy, so far as the system of a section of the Sakta Tantra is concerned, and should be appreciated from that point of view.

The systematisation of Tantric Philosophy, on its Sakta side, does not, seem to have yet been seriously attempted. The Sarva Darsana Sangraha of Madhavacharya ignores the Sakta School altogether. So do the other compendia, earlier and later. (*e.g.* Saddarsana Samuchchaya of Haribhadra, Sarvasiddhanta Sangraha of Sankaracharya, etc.). There are different lines of Sakta culture still in existence and we have reason to believe that some at least of these have preserved the philosophical tradition. The literature associated with the third Mahavidya, named Sodasi or Tripura Sundari is very extensive and presents several interesting feature of Tantrik literature. It is possible to construct a regular philosophy of the school out of the materials available to us and in this work of reconstruction the present treatise will, it is hoped, prove to be substantially helpful.

The Tripurarahasya, which claims to treat of the secrets of the Tripura Culture in all its aspects, is said to consist of three sections-*viz.* Mahatmya, Jnana and Charya. The Jnana Khanda has been published at Benares in a new edition1. The Mahatmya Khanda, of which (as of the Jnana Khanda) we have an original manuscript in the Government Sanskrit College, Benares, was entrusted for publication to the publishers of the Chowkhambha Sanskrit Series, Benares. The third section is apparently lost, no trace of it having yet been found. It is an extensive work of which the first two sections contain 2163 and 6687 verses.

The work is attributed to one Haritayana and the commentary called Tatparyadipika is from the pen of one Dravida Sri Nivasa, son of Vaidyanatha Diksita, resident of the village of Mahapuskara in the Dravida country. The commentary was composed in 4932 Kali Era (1831 A. D.).

The book is in the form of a discourse delivered by Haritayana to Narada. This discourse professes to be a reproduction of the teachings of Dattatreya to Parasurama and claims to be based on personal realisation and reason.

The plan of the work may be thus summed up. Parasurama having heard Dattatreya's lectures on the greatness of the Supreme Deity Tripura Sundari as embodied in the Mahatmya Khanda expressed a desire to the Master to be enlightened on the methods of worship for propitiating the Goddess. He was subsequently initiated in due form into the mysteries of Tripura worship and practiced penances for 12 years, under instructions from his tutor, at a hermitage on the Mahendra Hill in the South.

In the course of his spiritual exercises the ultimate problems of life and reality began to trouble his mind, and being unable to reach a solution himself be betook himself to the feet of his Guru for light and guidance. He has already heard a teaching on the subject from Samvarta, the great Avadhuta, on his discomfiture by Rama Chandra in the Treta age, but he had not been able to realise it at the time. He requested the master to explain to him the secrets of Samvarta's lessons, so that his doubts might be dispelled for ever. Dattatreya's response to Parasurama's question, interspersed with Parasurama's cross-questions here and there, constitutes the body of the Jnana Khanda. It would thus appear that the text of the Tripura Rahasya, which embodies the teachings of Haritayana to Narada, represents an old traditional lore of secret science originally revealed by Samvarta and subsequently expounded by Dattatreya to Parasurama.

The name Parasurama stands eminent among the votaries of Tripura. The Kalpa Sutra, treating of the secrets of the Tripura worship in ten Khandas and containing 335 Sutras, is attributed to Parasurama who is described as a scion of the family of Bhrgu, a disciple of Shiva and the son of Jamadagni and Renuka. This description of Parasurama implies that the Parasurama of the Tripura Rahasya is believed to be identical with the author of the Kalpa Sutras, though Pandit A. Mahadeva Shastri has, perhaps rightly questioned this identity. The tradition has it that Dattatreya was the author of a Samhita work in 18000 verses which was known under his own name (*viz.* Datta Samhita). Parasurama studied the extensive work and, to bring its contents within easy reach of students, summarised it in a body of Sutras distributed into 50 sections (Khandas). This contained 6000 Sutras, The Samhita and the Sutra were both summed up, in the form of a dialogue between Dattatreya and Parasurama, by Sumedha (pupil of Parasurama). This tradition is found recorded in the Tripurarahasya, Mahatmya Khanda.

It is evident that the work of Sumedha, who was of the Harita family and consequently known as Haritayana, is really to be identified with the Tripurarahasya itself rather than with the Kalpasutras of Parasurama as Laksmana Ranade has done, because the Parasurama Kalpa Sutra is not in the shape of a dialogue between Datta and Parasurama and is not attributed to Sumedha, whereas the Tripurarahasya has the form of a similar dialogue) and is ascribed to Sumedha Haritayana.

The line of Tripura worship is represented by several teachers. We have already referred to Dattatreya, and Parasurama. The names of Durvasas, Agastya, Lopamudra and several others may be added in this connection. Durvasas is associated with the authorship of a mahimnah stotra of the

Goddess, where he is described in the colophon as sakalagamacharya. Nityananda, who wrote a commentary on the above Stotra, says that Durvasas inter alia, Krodha Bhattaraka is really identical with Shiva Himself, who is the Master of the teachers of all the Agamas (sakalagama-charyachakravatim), as born from the womb of Anurapa.

The Supreme Goddess is variously named-as Tripura, Sundari Lalita, Sodasi, Sri Vidya, Kamesvari, etc. She is called Tripura, in as much as Her Body consists of three Saktis, *viz.* Brahma, Vaishnavi and Raudri. The Tripurarahasya speaks of Her in the following terms:

tripuranantashaktyaikyarupini sarvasaxini.
sa chitih sarvatah purna parichchedavivarjanath..

The partial appearance of the Self as thus occurring is known as bahyavabhasa, because such appearance implies the manifestation of what may be described as empty space which is other than the Self. Remembering that Chaitanya is all-embracing and can have nothing outside it-for if there were any such thing it would not shine out and would therefore be non-existent-what is popularly called the external is indeed only a reflection on Chaitanya as on a mirror. When the universe comes into being it does so as only an image within the unique Self. The universe as such is varied but underlying it is the pure and simple unity of Chaitanya revealing itself to the eye of diligent search (anusandhana). The manifestation of the universe, due to the Free Will (svatantrya) of the Absolute, is thus a process of Abhasa,-and for the initiation of this process nothing beyond the play of the Will is needed. The material and efficient causes, supposed to be necessary for every product, are held unnecessary.

The peculiar metaphysical position of the Tantra consists in the theory of Abhasa, which is consistent with this position. It rejects the Vivartavada of Neo-Vedanta, because the world is not originally a false appearance due to Error. It is real in the same way as an image is real, but it has no existence apart from the medium in which it is manifested. Its existence is only the existence of the medium. To the Vedantist the world appears as such to the ignorant owing to his ignorance and in the last analysis it is resolved into Maya which is not identical with Brahman and is material; but to a Tantrist the world is real and is expression of the Chit Shakti or Free Will of the Lord and is really spiritual in essence like the Lord Himself. In the last resort it turns back into the Chit Shakti which is never withdrawn, for the Will (svatantrya) remains, even after the world has disappeared.

The Vedanta system has had to fall back on the doctrine of Vivarta, because it denies in a sense svatantrya to Pure Chaitanya. The first stadium of creation is thus an Abhasa. The second stage which represents the subsequent condition shows how the Chit Shakti, already appearing (Abhasamana) in the Pure Chaitanya, further progresses. Maya emerges on the scene now and the Vivarta is the logical outcome. The third stage marks how Maya becomes productive. This is the Parinama or Evolution which gets on till the bhutas spring into manifestation. The fourth stage which represents creation out of the bhutas is known as Arambha or physico-chemical process of genesis. From the supreme stand-point of Tantra, however, the entire Creation is an Abhasa.

As thus realised She is the Eternal and supreme Truth beyond all limitations consequent on time and place. She is the essence of Chaitanya and is called Lalita owing to Her transcendent charms. The Shakti Sangama Tantra observes that it is this Lalita which assumes the form of Krsna as Purusa.

Sundari is one of the ten Mahavidyas (Mundamala Tantra, Patala 1). It is said (Ibid) that the ten Vidyas combined form a Mahavidya, but Sodasi is a Mahavidya by Herself. The Todala Tantra (Patala 1) calls Maha Tripura Sundari by the name of Panchami with Shiva (five-faced) as Her Bhairava. The Shakti Sangama however makes Lalitesvara Her Bhairava. This is different from Tripura Bhairava (or Vikarala, the companion of Chhinna) and Ghora Bhairava (*i.e.* Kala Bhairava, the companion of Dhumavati).

The system teaches that the Supreme Reality is of the nature of Pure Intelligence, which is self-luminous and unaffected by the limitations of time, space and causality. It has absolute freedom (purna svatantrya) in as much as its Power or Will (samkalpa) is unrestricted. This Power is really identical with the Essence of Chaitanya and remains either involved in it or expresses itself as its inalienable property. In the technical phraseology of the Shastra it is known as Vimarsa or Krpa, and is an eternal attribute of Chaitanya. The freedom referred to above implies that the Essence of Consciousness is free from vikalpas and is fundamentally distinct from matter. The Chaitanya is free, as it does not depend on anything else for its own revelation of matter.

The Power exists in a two-fold condition. What is generally known as creation or dissolution is in reality consequent on the manifestation of this Power or on its abeyance. It always functions, but its function is sometimes (*e.g.* during the creative period) expressed as the manifestation of the Universe till now absorbed in and identified with the Essence of Reality and at other times expressed as self-manifestation alone.

The Supreme Reality of the Agamas would thus seem to differ in a sense from the Brahman of Vedanta. Though both are essentially of the nature of Intelligence there is a fundamental distinction between the two. The Absolute of the Tantra is endowed with Power which is held to be identical with Itself and by virtue of which It is described as the Free Agent (svatantrakartta). Freedom to act forms the essence of Chaitanya. In other words, according to the Tantric viewpoint, Shiva and Shakti are aspects of one and the same Reality. But in the current non-dualistic school of Vedanta Brahman, which as in this Tantra is described as of the nature of Pure Consciousness, is no better than an action-less Locus (adhikarana), on which the Power, which is attached to It mysteriously and is neither identical with nor distinct from It, plays.

It is conceived as a Pitha or passive background in relation to the active power operating on It. The Shakti, called Maya in the Vedanta School, is not thus of the nature of Brahman but is material (anirvachaniya), though it is held to be, of course mysteriously, subservient to it. But as conceived in the

Tantra Shakti or Pure Freedom is absolutely non-material. The term Chit Shakti used to denote this power implies its spiritual essence.

What in the Tantras is known as vahyabhasa or the manifestation of a non-ego (anahambhava) within the Pure Ego (shuddhatma) but appearing as external to it is tha Radical Nescience (mulavidya) of Vedanta. This non-ego is the so called Avyakta (Unmanifest) or Jada Shakti (Matter). But the Freedom or the Spiritual Power (Chit Shakti) of the Lord, as described in the Tantras, is beyond the Nescience referred to above, and to this Power the Advaita Vedanta seems to be a stranger.

In as much as the Avidya itself or the Material Power is a product of the Spiritual Power which is the ultimate source of all existence there is no discrepancy in the statement, often found in Tantric Literature, that this Power has three distinct states of its existence:-

(a) During the universal dissolution when the Self is free from all vikalpas the Shakti exists as Pure Chit Shakti or Chit Prakrti.

(b) When the vikalpas are on the point of merging-when though there is no vikalpa as such there is yet a tendency in the direction of vikalpas-the Shakti is called Maya Shakti or Jada Prakrti.

(c) But when the vikalpas are fully developed and materiality becomes dense the Shakti appears as Avidya.

It has already been observed that the appearance of the universe follows upon the self-expression of the Divine Power and the Cosmic End follows from the withdrawal of the self-same Power.

After the period of Cosmic Night is over the Will of the Lord, in co-operation with the mature adrsta of Jivas, manifests only partially, as it were, the Essence of the Self, whereby the Self is revealed as limited.

The appearance of limitation is thus the emergence of not-self, known as Avidya or Jada Shakti, called also by the name of Void (Shunya), or Prakriti or Absolute Negation or Darkness (tamah) or Akasa. This is the first stage in the order of creation and represents the first limitation imposed on the Limitless. The erroneous belief, generated through the Freedom of the Lord-the Self-that the Ego is partial (ekadeshika) and not full and universal (purna) is responsible for the appearance of this Something which being a portion of the Self is yet outside of it and free from self-consciousness and is described as not-self or by any other name as shown above.

Thus the Supreme Reality splits itself spontaneously, as it were, into two sections-one appearing as the subject and the other as the object. The Purnahanta which is the essence of Supreme Reality disappears after this cleavage: the portion to which limited egoism attaches being the subject and the other portion free from egoism the object. The object as thus making its appearance is the Unmanifest (avyakta) Nature from which the entire Creation emanates and which is perceived by the subject as distinct from itself.

It has been observed that Chaitanya is of the nature of self-luminous Light (sphurath prakasha), which may shine on itself (svatma), in which case it is known as Ahanta, or I-ness may rest on the Non-ego (anatma) and express itself as Idanta or This-ness. The essence of Chaitanya consists in the fact that the light (prakasha) is always confined to itself. This universal Ego or 'I' stands, behind all dualism. The Supreme Ego is universal, as there is nothing to limit (parichcheda) or to differentiate (vyavritti) it, and the entire visible universe exists in identity with it. But this characteristic by its very nature is absent from Matter (Jada), which is not self-manifest. Just as light and heat coexist in fire, in the same way universal Ahanta and Freedom or Shakti coexist in Chaitanya. This freedom is Maya which though essentially identical with Chaitanya (chidekarupa) brings out varieties of an infinite kind, but in bringing out this variety it does not in the least swerve from the Essence.

The appearance of the Universe in Pure Chaitanya is the action of Avidya, which has three distinct stages:

(a) The first is the germinal state (bijavastha), when the material power, which is still in its earliest phase of manifestation, is pure. Matter does not assert itself at this stage and consequently there is no differentiation in experience. In other words, it does not yet appear as distinct from Chaitanya, though potentially it exists. This stage is represented by the five pure Tattvas, *viz.*, Shiva, Shakti, Sadashiva, Suddha vidya and Isvara.

(i) The Avidya, which has been described above as being the Chaitanya in its limited appearance as an object external to the subject is called Shiva. In pure Chaitanya, owing to the play of Its own Will, an infinite number of limited aspects (spamsha) arises. These are mutually distinct. From this point of view to every limited aspect of Chit there is a corresponding object external to it (bahyabhasa), but to the Unlimited Chit or Pure Self (purnashratma = parashiva) there is no externality. The universal (samanya) common to all the pure and limited Chit aspects referred to above is called Shiva Tattva. This Tattva is thus a Samanya holding within it all the Visesas, but Para Shiva or Pure Self is transcendent and above both Samanya and Visesa. Hence Shiva Tattva may be more properly described as Pure Chaitanya in its general but conditioned form, free from all Vikalpas and is to be distinguished from the Absolute proper.

(ii) The appearance of Shiva (parichchinna nirvikalpachith) as ahamh is called Shakti. Although this self-presentative character (ahambhasana) is in the essence of Chit, so that there can be in fact no differentiation between Shiva and Shakti as such, the Chit is nevertheless known as Shiva in so far as it is free from all visesas and as Shakti by virtue of its characteristic self-awareness (ahambhasana).

(iii) When the self-presentation (ahambhasana) is no longer confined to the Self but is extended to the not-self or the object (mahashunya) external to the Self it is known as Sadashiva. This state marks the identification of the Self with the not-self in the form "ahameva idamh" and indicates predominance of spirit over matter.

(iv) But when matter prevails and the consciousness assumes the form "idamh ahamh" the state is technically called Isvara.

(v) The term Suddha Vidya is reserved for the state which represents an equality in the presentation of the subjective and objective elements in consciousness.

(b) The second stage in the evolution of Avidya, described as ankuravastha, represents a further development of difference or materiality, when the subtle products of matter and spirit make their appearance. In this mixed condition both spirit and matter are equally predominant and the seven mixed (mishra) tattvas, viz, Maya, Kala, Vidya, Raga, Kala and Niyati reveal themselves.

(i) The confirmation of difference due to the Free Will of the Supreme, which characterises the second stage, has the effect of reversing the normal relation between spirit and matter. Thus while in the first stage described above Spirit or Chit Shakti dominates matter or Jada Shakti which exists in a rudimentary state, merged in spirit or Self, the second stage shows the preponderance of matter over spirit. Consciousness loses its supremacy and becomes a quality inherent in the material subject. All this is due to the emergence and development of bhedasamkalpa in Chaitanya. This material subject-which is matter prevailing over spirit and related to it as a substance to its quality-is called Maya.

(ii-vi) The five aspects of Maya are the five so-called Kanchukas which are the five eternal Saktis of Para Shiva in a limited form. The obscuring power of Maya acts is a veil as it were upon the Omnipotence, Omniscience, Self-contentment, Eternity and Freedom of the Supreme Self and thus acting is known as Kala, Vidya, Raga, Kala and Niyati respectively.

(vii) The Pure Self as obscured by Maya and its fivefold activities appears as Purusha with its limitations of action, knowledge, contentment, eternity and freedom.

(c) The third or grossest stage in the evolution of Avidya is represented by the dense products of the mixed tattvas, where matter is overwhelmingly strong. This stands for the group of the twenty four tattvas, from the Primary Prakriti down to Prithivi, constituting the material order.

Prakriti, with which the lower creation begins, is indeed the assemblage (samashti) of the Vasanas of all persons with various and beginningless Karmans: it may be fitly described as the body of the Karman Samskaras of the Jivas, considered as inhering in Chit Shakti or Self. This Karma vasana or Prakriti is threefold according as the experience which is the moral outcome of this vasana is pleasant or painful or of the nature of a comatose condition in which neither pleasure nor pain is felt.

The Vasanas exist in a twofold condition, as Avyakta when they lie unmanifest in dreamless sleep or as Chitta when they manifest themselves in dreams and wakeful states. In the dreamless state there can be no experience of pleasure and pain, because the mature Karmans having been worked off through experience the others which are not yet ripe are not ready for fructification.

It is a fact that Karmans, when they are matured by time, cause the Jnana Shakti of the Conscious Self to move outwards and have contact with the objective world. In a state of sleep such movement is naturally absent. But the process of time during which the sleep continues acts on the Karmans and matures some of them, so that the Jnana Shakti is allowed to come in touch with the external objects or with their ambulances and sleep is over. The Shakti as thus qualified by the body of

Karma-Vasanas leading to contact with the objects and consequent enjoyment (bhoga) is known as Chitta.

The Chitta differs according to the difference of Purusa but it is one with Prakriti in dreamless sleep. Thus the Chitta may be viewed as Purusha or as Prakriti according as the conscious (chiti) or unconscious (avyakta) element prevails in it. It is not, therefore, a distinct category, but falls either under Purusa or under Prakriti.

Lalita's Paradise Island

Aim Hrim Shrim hail to the sea of nectar; island of Gems; garden of Various Trees; Aeon Tree garden; Santana garden; Hari Chandana garden; Mandata garden; Parijata garden; Kadamba garden; Enclosing wall of Topaz; Enclosing wall of Ruby; enclosing wall of Gomeda gem; enclosing wall of Diamond; Enclosing wall of Vaidurya gem.

Aim Hrim Shrim Hail to the Enclosing Wall of Sapphire; Enclosing Wall of Pearl; Enclosing wall of Emerald; Enclosing wall of Coral; Jewelled Pavilion; Pavilion Adorned with 1000 Bunches of Flowers; nectar Lake; Bliss Lake; Mirror Lake; Shining sun; Shining moon; Great Beautiful Gate; Great Forest of Lotuses; Palace made of Wish Fulfilling gem.

Aim Hrim Shrim Hail to the Eastern Face Doer; Southern Face Door; Western Face Door; Northern Face Door; Circle of Jewelled Lamps; Great Lion Throne Made of Gems; Brahma Couch Foot; Vishnu Couch Foot; Rudra Couch Foot; Ishvara Couch Foot; Sadashiva Couch Foot; Hamsa Mattress; Pillow on the Hamsa Mattress; Mattress strewn with safflower; the great canopy; Aim Hrim Shrim hail to the great curtain.

One should worship using these 44 temple mantras, employing flowers and rice.-Nityotsava, translated in Lalita Magic, Sothis Weirdglow 1983.

There is a beautiful island, scented by fragrant breezes, with a multitude of gardens, lakes and bowers and there Shri Shri Lalita Devi resides. Before you enter through the great curtain you must pass through four gates, to the east, south, west and north of Her lion throne. Her canopy is also a hidden gate. This symbolism has some similarities to Shiva with his five faces. The scheme is set out in full below for the lovers of intricacy.

East of the Lion Throne

(1) *Bala*: Her mantra is Aim Klim Sauh. She dwells in the centre of a forest of trees with bright orange-coloured blossoms (Kadambas), in the centre of a Jewelled pavilion, below a great Aeon Tree, seated on a Jewelled throne, a priceless gem as Her diadem, at Her feet a Jewelled pot, with jewels as necklace, effulgent bracelets, holding a book and dispelling fear with Her left hand, with Her right holding a rosary,

(2) *Sampatprada Bhairavi*: Mantra: Hsraim Hasakalarim Hsraum. She is as bright as 1000 rising suns, charming, wearing red linen garments, with a garland of flowers on her head, her breasts smeared with red unguent, carrying a china rose, a hook, dispelling fear and granting boons, with 3 eyes, a face full of love-play, the Moon as her diadem, smiling slowly.

(3) *Chaitanya Bhairavi*: Mantra: Shaim Sa Ka La Hrim Shrauh. Like 1000 rising suns, adorned with various Jewels, crescent moon as diadem, wearing red clothes, carrying noose and goad, bestowing boons and dispelling fear, she is adorned with skulls, has massive round swelling firm breasts.

(4) Chaitanya Bhairavi (b). Mantra: Shaim Sa Ka La Hrim Nityaklinne Madadrave Sauh. She carries noose and goad, and a skull, she is red, garlanded with skulls, seated on a throne made of five corpses.

(5) *Kameshvari Bhairavi*: Mantra: Ha Sa Kha Phrem Ba Sa Ka La Rim Ha Saum. She is bright as 1000 rising suns, the Moon is her diadem, she has 3 eyes, beautiful gems, red clothes, garland of bloody heads, carries a trident, a damaru, a sword, a shield, a bow and arrow, a noose and a goad.

South of the Lion Throne

(1) *Aghora Bhairavi*: Mantra: Aghore Aim Ghore Hrim Sarvatah Sarvasarvebhy Ghoraghoratare Shrim Namaste astu Rudrarupebhyah Klim Sauh. She is reddish, like a 1000 suns, the Moon is bound in her hair, she wears various gems, has pearls as a diadem, wears a blossoming red line of skulls, has 3 beautiful eyes, with large swelling breasts, wearing red clothes, young, wanton, holds a book and dispels fear in her two left hands, and in her right holds a rosary and bestows boons. She is seated on a great corpse.

(2) *Mahabhairavi*: Mantra: Ha Sraim Ha Sa Ka La Hrim Ha Srau. Description as (1) above.

(3) *Lalitabhairavi*: Mantra: Hrim Klim Hsraum. She is like 1000 rising suns, with flaming rubies as her diadem, red pearl earrings, limbs smeared with red unguent, bedecked with beautiful red flowers, wears red clothes, holds noose, goad, book and rosary. She is the Kumari form.

(4) *Kameshi Bhairavi*: Mantra: Klim. She is effulgent as the China Rose. She carries bow and arrows and is adorned with precious gems.

(5) *Raktanetra Bhairavi*: Mantra: Saim Sa Ka La Rim Sauh. She has a flaming ruby coloured diadem, red pearl earrings, her limbs all smeared with red paste, a garland of beautiful red flowers, red clothes, holds a noose, goad, book and rosary, is wanton, youthful, without shame, has beautiful rising large breasts, beautiful buttocks and lovely slender waist.

West of the Lion Throne

(1) *Satkuta Bhairavi*: Mantra: Ha Sa Ka La Ra Daim Ha Sa Ka La Ra Dim Ha Sa Ka La Ra Dauh. She is as brilliant as the sun at dawn, resembling the China Rose, wearing a garland of skulls, at her feet is a gold vessel, she has large swelling breasts, holds a noose, goad, book and rosary. Sits on a corpse.

(2) *Nitya Bhairavi*: Mantra: Da Ra La Ka Sa Haim Da Ra La Ka Sa Him Da Ra La Ka Sa Hauh. Seated on 5 skulls, wears a garland of skulls, as effulgent as the sun. Adorned with gems, holds a noose and goad and dispels fear and grants boons.

(3) *Mrita Samjivani*: Mantra: Hrim Hamsah Samjivani Jum Jivah Pranagranthim Kuru Kuru Kuru Kuru Svaha. No description given.

(4) *Mrityunjayapara*: Mantra: Vada Vada Vagvadini Ha Saim Klinne Kledini Mahakshobham Kuru Kuru Hasaim Om Moksha Kuru Kuru Ha Sauh. Dwelling in the centre of the Kadamba Forest, she holds a book in her left hand, and in her right a rosary. She is as white as dazzling jasmine.

(5) *Vajra Prastarini*: Mantra: Hrim Klinne Aim Krom Nitya Madadrave Hrim. She is in the centre of the Kadamba Forest, red, with a beautiful crescent Moon (Moon Kala) as her diadem, very red, with 3 beautiful eyes, upon a great yantra in the centre of a crimson sea, holds pomegranate, arrow, noose, goad, bow, skull. She is smeared with red.

North of the Lion Throne

(1) *Bhuvaneshvari Bhairavi*: Mantra: Hasraim Ha Sa Ka La Hrim Hrasaum. She is as brilliant as a China Rose, like a pomegranate flower, with a crescent moon as her diadem, 3 eyes, red clothes, various beautiful gems, large rising swelling breasts, seated on a corpse, with a garland of skulls, holds a noose and goad and grants boons and dispels fear.

(2) *Kamaleshvari Bhairavi*: Mantra: Sahraim Sahakalahrim Sahraum. Description as (1) above.

(3) *Siddha Kaulesha Bhairavi*: Mantra: Hasraim Hasrim Hasrauh. She is a Kumari, effulgent as 1000 rising suns, with 3 eyes, Moon as her diadem, slender waisted, large buttocks, wearing large Jewels, adorned with red, wanton, youthful, proud, large high swelling breasts, wears a garland of skulls, beautifully bejewelled, holds

a book and dispels fear with her left hands, in her right she holds a rosary and grants boons. She sits on a corpse.

(4) *Damara Bhairavi*: Mantra: Hashaim Haklim Hasauh. Like a Bandhuka flower, seated on 5 skulls, the Moon is the peak of her beautifully gemmed diadem, wears a garland of skulls, has 3 eyes, red clothes, large rising swelling breasts, holds book, rosary, gives boons, dispels fears.

(5) *Kamini Bhairavi*: Mantra: Hasaim Ha Sa Klim Hasauh. She wears red jewels and ornaments. Holds a noose, a goad, grants boons and dispels fears, sits on a corpse, she is garlanded with skulls.

Great Upper Direction

(1) *Sundari*: Mantra: Haim Ha Ka La Hrim Hasauh.

(2) *Sundari*: Mantra: A Ha Saim Ah A Sim A Hasauh.

(3) *Sundari*: Mantra: Aim Ha Sa E Ha Sa Ha Shaim Ha Ha Ha Ka La Hrim Ha Ha Ha Harauh.

(4) *Sundari*: Mantra: Ka La Sa Sa Sa Sa Haim Ka La Ha Ha Sa Sa Sa Ha Haim Ka La Ha Ha Sa Sa Sa Hrim.

(5) *Sundari*: Mantra: Ha Sa La Ksha Sa Ha Sa Hauh Sa Ha Ha Sa La Kshah Saim Ha Sa Ha Sa La Ksha Ha Sim.

This concludes the description of the Red Yoginis who dwell in the five directions around Devi Lalita in Her palace of gems. The bija mantras, of tongue twisting perplexity have an inner logic related to the root mantra.

The Shri Vidya Ratna Sutras shows that Devi Lalita outrays thousands upon millions of rays which are each Devis of a certain colour. The 15 Goddesses above are important in the overall scheme, and mentioned in such works as Dakshinamurti Samhita and Jnanarnava Tantra—both important tantras of the Shri Vidya path. A few of the mantras are Sanskrit words interspersed with bija—for example South 3—Speak! Speak! Giver of Speech! Hasaim. Moisten! O Moistener! Cause Great Agitation! Cause Great Agitation! Hasrim. Om Give, Give Liberation! Hasauh.

3

The Other Traditions of Tantra

Devirahasya Tantra

There are many compilations or tantrik digests, discussing a variety of topics a sadhaka or sadhvini needs to know. Below is a summary of the contents of the *Devirahasya*, which will give an idea of the scope of this type of work. There is little philosophy here; practically the entire contents of the work deals with mantra, yantra, puja and sadhana of the different gods and goddesses discussed.

Rahasya means secret and the work does cover most of the topics a practitioner would need to know. These include purashcarana, which is the preparatory work before puja proper can start. This is very arduous, involving the recitation of mantra and a ritual which spans many hours. The *Devirahasya*, however, introduces some short-cuts for the Kaula initiate.

The panchangas (five limbs) in this work give essential puja information for a sadhaka or sadhvini and give intriguing insights into various aspects of devata worshipped by the Hindu tantrikas.

Chapter 1

This deals with the characteristics of guru and pupil, with the planetary positions and times of initiation, and with the attendant disqualifications on both pupils and gurus. It also deals with the sequence of initiation, the purification of the disciple, and the initiation of shaktis.

Chapter 2

The different mantras of the Devis are given. The Devis mentioned are Bala (an aspect of Tripurasundari as a young girl), Panchadashi and Sodasi (Tripurasundari), Tripura, Vidyarajni (Queen of Vidya), Bhadrakali, Matangi, Bhuvaneshvari, Ugratara, Chinnamasta, Sumukhi (Pretty Face), Saraswati, Annapurna (Full of Food), Mahalaksmi, Sarika (Small Bodied), Sarada (Autumnal), Indraksi (Indra's eyes), Bagalamukhi, Mahaturi (The Transcendent Fourth), Maharajni (Great Queen), Jvalamukhi (Fire-Mouth), Bhida, Kalaratri (Night of Time), Bhavani, Vajrayogini, Dhumravarahi(Smoky Varahi Devi), Siddhalakshmi, Kulavagisvari, Padmavati, Kubjika (Crooked One), Gauri (Fair One), Khecari, Nilasarasvati, Parasakti.

Chapter 3

Gives the different Shaiva mantras such as: Mrtyunjaya (Shiva as Conqueror of Death), Amrtesvara, Vatukabhairava (Shiva in His aspect as a terrifying boy), Maheshvara, Shiva, Sadashiva, Rudra, Mahadeva, Karala(Formidable One), Vikarala, Nilakantha, Sarva, Pasupati (Lord of Beasts), Mrda, Pinaki, Girisa, Bhima, Mahaganapati, Kumara, Krodhanesa, Isa, Kapalisa, Krurabhairava (Cruel Bhairava), Samharabhairava (Dissolution Bhairava), Isvara, Bharga, Rurubhairava, Kalagnibhairava, Sadyojata (instantly arising—a name of Shiva as penis), Aghora, Mahakala and Kamesvara.

Chapter 4

The different mantras of Vishnu are here given. These are the Lakshmi-Narayana mantra, and the mantras of Radhakrsna, Vishnu, Lakshmi-Nrsimha, Lakshmi-Varaha, Bhargava, Sita-Rama, Janardana, Visvaksena and Lakshmi-Vasudeva;

Chapter 5

The different Utkelana (laying open) of the mantras given above are given. These are mantras which themselves open the mantras up to use.

Chapter 6

Gives the vitalising mantras of the Devatas described in chapters 2,3 and 4.

Chapter 7

In this chapter the mantras used for reminding any curses that may have become attached to the mantras in chapters 2, 3 and 4 are given.

Chapter 8

The method of reciting mantras is here described. The guru puja mantra is given.

Chapter 9

Deals with the method of putting together the mantras already described in chapters 2,3 and 4.

Chapter 10

Purascarana, or the performance of acts by which a given mantra may be made efficacious, is described in this chapter. This is performed by reciting it 400,000, 200,000 or 100,000 times. It should be performed under a fig tree, in the wilderness, in the cremation ground, in a desert, at crossroads, and should be started at midnight or mid-day. Purascarana should be done under auspicious astrological configurations after having worshipped one's own guru. A yantra is described which should be used in its application.

The sadhaka has to fill four pots at the cardinal points. At the end of the chapter alternative methods of doing this necessary act are described. These are through sexual intercourse with an initiated shakti, by reciting the mantra during the birth of a child of the in-group, on a dead body in a cremation ground, during the time the Sun takes to rise and set, in a solar or in a lunar eclipse.

Chapter 11

Continues the topic of the previous chapter, and describes the homa which should be done.

Chapter 12

Describes in code form the unfolding of the different yantras of the Devatas described in chapters 2,3 and 4.

Chapter 13

This chapter describes how an amulet (kavacha) may be made of the yantra of one's own Istadevata, bound into a ball, and carried upon the person. This amulet is said to give miraculous results. The yantra should be drawn upon birch-bark using 8 different kinds of scent. These are described as svayambhu, kundagola, Rocana, Aguru, camphor, musk, honey, and that arising from Malaya (*i.e.* sandal). The first two are well-known in the tantras as arising from various Kula women at the menstruation time. The others have similar significance in the left handed and Kaula tantras. Various methods of purification are given in the text, and it is declared that the 1,000 names of the particular Devata should be written around the yantra.

Chapter 14

Gives details of the Rishis or seers of the various mantras.

Chapter 15

The sadhana of the cremation ground. begins to be described. This chapter contains only 13 verses but there is an extensive commentary provided.

Chapter 16

The different Bhairavas of the elements have to be worshipped. Mahakala-bhairava is the Seer of the mantra, Ushnik is the metre, Sri Smasana is the Devata, Hrim is the bija, Hum is the shakti and Krim is the kilaka. The application of the mantra is in the attainment of the four aims of mankind.

Chapter 17

Purification of the rosary formed from human skulls is discussed here.

Chapter 18

In this chapter rosary and yantra purification is dealt with.

The nature of the five products of the cow and the Yantresvari mantra are also discussed.

Chapter 19

The origin of wine is the subject of this chapter. Nine vessels which form the receptacles in which wine is kept are discussed. The presiding Devatas of these are Sadashiva, Isvara, Rudra, Vishnu, Paramesti, Indra, Guru,(Jupiter), Sukra (Venus) and the Sun and the Moon taken together.

Chapter 20

Gives further details concerning wine.

Chapter 21

The Santi Stotra commences this brief chapter. This hymn removes the curse attached to wine.

Chapter 22

This continues the topic of wine, and discusses how the same may be purified. It gives details concerning Anandabhairava and his Shakti Suradevi. The gayatri of the former is given as: Anandesvaraya vidmahe Sri Suradevyai Dhimahi tanna Ardhanarisvara pracodayat (Let us think of the Lord of Bliss, let us contemplate the Auspicious Suradevi. May that half-Shiva and half-Shakti form direct us). The dhyana of Tiraskarani Devi is given towards the end of the chapter together with Her prayogas (rituals), etc. She confers invisibility on a sadhaka.

Chapter 23

This chapter deals with the purification of nine Saktis, who are Nati (actress), Kapaliki(bearing skulls), Vesya (whore), Rajaki (washer-woman), Napitangana (barber's wife), Brahmini, Sudrakanya (Sudra's daughter), Gopalakanyaka (Cowherd's daughter) and Malakarakanya (Daughter of a Garland-Maker). The Devata of this rite to follow is called Parambika.

The best time for the rite is at midnight. The puja sequence is given, and it is stated that the girl should be placed on the left of the sadhaka in a Sri Cakra. She has to have dishevelled hair, be free from shame, and adorned with jewels. The various mantras of each of these nine Kumaris are given. Details are given of the left handed cakra of eight or eleven couples and the mantras to be used.

Chapter 24

This gives various materials of which a rosary may be made, as also the way knots and so on are to be tied. The rosary made of human skulls is described, as well as rosaries made from various trees, tulasi rosaries, crystal, rudraksha, jewels, gold, and rosaries made from lotus seeds and human teeth. The last, and the first, are to receive specific kinds of purification.

Chapter 25

The purification of yantras and the various materials from which they are made are discussed in detail. Yantras are spoken of as eightfold as being made from gold, silver, copper, crystal, birch, bone, hide and Vishnu-stone. The mantras for purifying these eight materials are given, and rites performed at night are discussed at the end of the chapter.

Chapters 26-30

These chapters comprise the Ganapati Pancangam (5 limbs). These limbs are (i) Mantra, Yantra, Dhyana and the six karmas or magical acts (ii) The worship of Mahaganapati (iii) The Kavaca or Armour of Mahaganapati (iv) The 1000 Names of Mahaganapati and (v) the Mahaganapati Stotra or Hymn.

Chapters 31-35

The Surya Pancangam or the Five Limbs of the Sun.

Chapters 36-40

The 5 limbs of Lakshmi-Narayana.

Chapters 41-45

The 5 limbs of Mrtyunjaya.

Chapters 46-50

The 5 limbs of Durga Devi.

Chapter 51

Deals with the rahasya or secret of Durga Devi.

Chapter 52

This brief chapter describes the mantra sadhana of the Durgarahasya, deals with enlivening of the mantra, and with its putting together.

Chapter 53

Discusses Nilakantha, or the blue-throated manifestation of Shiva, when he drunk the poison produced by the churning of the Milk Ocean. It gives his mantra, dhyana, and the seer.

Chapter 54

Discusses initiation, and its time, and deals with Guru initiation..

Chapter 55

This chapter discusses purashcharana and deals with the same done for the disciple by the Guru. It describes, in addition, the best places for doing the same.

Chapter 56

Pancaratnesvari, or the Devi of the five jewels, together with the mantra unfolding of Durga, Sarada, Sarika, Sumukhi and Bagala are dealt with here.

Chapter 57

Homa done at night in the cremation ground forms the substance of this section.

Chapter 58

Deals with the characteristics of cakra worship, the nature of those sadhakas entitled to it, the placing of the pot (kumbha) and the giving of bali or animal sacrifice. At the end Kanyapuja, or worship of virgins, is described.

Chapter 59

This chapter discusses the different paths, such as Daksinacara, Vamacara, and Kulacara.

Chapter 60

Deals with the guru.

Jnanasankalini Tantra

Hindu tantras are discourses between Shiva and Shakti, the male and female aspects of divinity whose play creates the entire universe. The *Jnanasankalini Tantra* is a brief work of 107 shlokas (verses) which outlines the dynamics of this interplay.

Of particular importance in this short work is the emphasis placed on the syllable Om, made up of the three Sanskrit letters a+u+m. These represent Shiva, Shakti and their union and can also be represented by the three gunas or qualities well known as rajas, tamas and sattvas.

Other important elements of the tantrik cosmology are outlined here, including the correspondence between the macrocosm and the microcosm, the five elements of earth, air, fire, water and space, and the essential similarity between the individual spirit, the Atma, and the universal spirit, the Paramatma.

The emphasis here is on Jnana, or pure knowledge. Although the spirit is one and all-pervading, it manifests through a variety of elements (tattvas). Through ignorance, an individual soul (jiva), may take each or any of these elements to be himself or herself.

The work here translated is, then, a brief summary of the essential elements of the Hindu tantrik tradition. There is no indication in the contents of this book when it was written, but it cannot be very old.

Translation

Mahadevi questioned Maheshvara, the Deva of all devas, the World Guru, sitting on the peak of Mount Kailasha: Speak to me of knowledge!

Devi said-Deva! what is creation and how is creation destroyed? What is the Brahma Jnana beyond creation and destruction?

Ishvara answered-Creation comes from that beyond words (avyakta) and returns to that beyond words at (the time of) destruction. Brahma Jnana is the avyakta, beyond creation and destruction.

From the syllable Om comes everything, the fourteen vidyas, mantra, puja, dhyana, action and non-action.

The four vedas, the six limbs of veda, mimamsa, nyaya, dharma shastra and puranas are the fourteen Vidyas.

For as long as these Vidyas are known, (true) knowledge is not. On knowing Brahma Jnana one is strong in all other knowledge.

The vedas and the puranas are like common whores but the Shambhavi Vidya is concealed like a Kula Shakti.

All knowledge is in the body, all devatas are in the body, all sacred bathing spots are in the body, known through the words of a guru.

Self-knowledge is the cause of happiness and liberation for a man, not dharma, not karma and not the reciting of mantra.

Just as there is potential for fire in wood, as there is scent in flower and nectar in water, so in the centre of the human frame is Deva, devoid of good or bad.

The Ida is the Ganges and the Pingala is the Yamuna. In the centre of the Ida and the Pingala is the Sushumna, the Saraswati.

The union of the three is known as the king of bathing places. Whosoever bathes there is released from all badness.

Devi said-Maheshvara, what is Khecari Mudra? What is Shambhavi Vidya? What is knowledge of the self? Tell me of that!

Ishvara said-A steady mind without external support, steady breath without restraint, steady seeing without looking, that resembles Khecari Mudra.

Like a child or a fool's mind drifts in sleep and dream, even without actually being asleep, so a man who wanders similarly without support knows Shambhavi Vidya.

Devi said-Devadeva, Natha of the Cosmos, Parameshvara, tell me the differences between the different darshanas, each separately.

Ishvara said-The tridandins are devoted to practice of the vedas. The Shaktas worship Prakriti, the Buddhists follow the void. (Shunya).

The Carvakas, although knowing the tattvas, are atheistic prattlers, denying the existence of the source of all.

Uma said, Lord! Speak of the characteristics of the microcosm! Talk of the five elements and the twenty-five qualities.

Ishvara said-The five qualities of earth (Prithvi) according to Brahma Jnana are bone, flesh, nails, skin and hair.

Knowers of Brahma Jnana say that semen, blood, marrow, urine and faeces are the five gunas of water.

The knowers of Brahma Jnana say sleep, hunger, thirst, fatigue and idleness are the five qualities of fire.

According to Brahma Jnana, holding, shaking, throwing, expanding and contracting are the five qualities of air.

The knowers of Brahma Jnana say desire, anger, delusion, shame and greed are the five qualities of space.

Air comes from space, fire comes from air, water comes from fire, and earth comes from water.

Earth dissolves into water, water dissolves in fire, fire dissolves in air and air dissolves in space.

Creation comes from the five tattvas and is dissolved into them. Greater than the five tattvas is that which is above them, without stain.

The organs of touch, taste, smell, sight and hearing are the five tattvas of the senses. Mind (manas) is the tattva that gives birth to them.

The entire macrocosm is situated in the centre of the body. That with form is destroyed while that without form is indestructible.

Whosever's mind is fixed on the formless becomes one with the formless. Therefore, by every means, reject that with form.

Devi said-Adi Natha, tell me of the seven dhatus. What are they? What is Atma? What is the inner Atma (Antaratma)? What is the Supreme Atma (Paramatma)?

Ishvara said-Semen, blood, marrow, fat, flesh, bones and skin are the seven (dhatus) in the body.

The body is Atma. Mind is the Antaratma and Paramatma is the Shunya (void) where the mind dissolves.

Blood is the mother dhatu and semen is the father dhatu. The Shunya dhatu produced from vital breath is the foetus.

Devi said-Tell me of the nature of speech. How is it produced and how is it dissolved?

Ishvara said-Vital breath comes from the Avyakta, mind arises from vital breath. Speech comes from mind and is dissolved into mind.

Devi said-Where does the sun dwell? Where does the moon dwell? Where does vital breath dwell? Where does mind dwell?

Ishvara said-The moon is at the root of the palate. The sun is at the root of the navel. Vital breath is above the sun and mind is below the moon.

The Cit is above the sun and the jiva is above the moon. Mahadevi, this is to be learned from the mouth of a guru.

Devi said-Where is Shakti? Where is Shiva? Where is Time? What causes ageing?

Ishvara said-Shakti is within (Patala). Shiva is without. Time is in space and causes ageing.

Devi said-What wants to eat? What wants to drink? What stays awake in waking, dream and deep sleep?

Shiva said-Vital breath wants food. Food is consumed by fire. Vital breath is awake during waking, dream and deep sleep.

Devi said-What causes karma? What causes bad deeds. Who does bad deeds? How 'is one liberated from bad deeds?

Shiva said-The mind does bad deeds. The mind is stained by bad deeds. That mind, having become itself, does neither good nor bad.

Devi said-How does an embodied soul (jiva) become Shiva? What is cause and effect? Speak to me, of your grace!

Ishvara said-Bound by delusion jiva, freed from delusion Sadashiva. You are cause and effect. Enlightenment is different from you.

Folk subject to tamasa wander to this sacred place and that sacred place, thinking that Shiva is here, Shakti is elsewhere, mind is somewhere else and vital breath is somewhere else again.

O Beauteous One, they do not realise the sacred place within the body. How then, can they be liberated?

Veda is not the vedas, the eternal Brahma is veda. Whosoever knows the Brahma Vidya is a Brahmin, skilled in the vedas.

After churning the four Vedas and all the scriptures, the yogis have extracted their essence, leaving pandits to drink the butter milk which remains.

All the scriptures are like dirty leavings with everyone speaking of knowledge. The Brahma Knowledge beyond words is not dirty leavings.

The path of Brahma is the ultimate tapas and not ordinary tapas. Whosoever knows the upper semen, he is a deva.

Meditation is not meditation unless the mind is united with Shunya. He who does so becomes happy and liberated, no doubt.

Sacrifice (homa) is not homa unless it is samadhi, in which the vital breaths are sacrificed in the fire of Brahma. That is the true sacrifice.

Good and evil acts give rise to good and evil consequences. Therefore, by every means, a wise man rejects them.

For as long as ignorance prevails, so long does caste and family matter. After knowing Brahma Jnana, one is free from all distinctions of caste.

Devi said-Shankara, I do not understand this knowledge you have spoken of. Devesha, remove my doubts! Tell me how the mind is dissolved.

Shankara said-It is said that Brahma Jnana is like dreamless sleep in which is dissolved mind, speech and actions.

Brahma Jnana, it is said, is a state of one-pointedness, without anxieties, peaceful, free of delusion and bestowing a child-like nature.

In the next line I will declare that spoken by the knowers of the truth. Yoga is the the rejection of all thoughts and concerns.

One who reaches samadhi for a split second or even half a split second destroys the evil of a hundred births instantly.

Devi said-Deva, what is Shakti and what is Shiva. Tell me about this! Reveal the nature of Jnana!

Shakti dwells in the moving mind and Shiva dwells in the steady mind. He who is established in the steady mind becomes accomplished while living in the body.

Devi said-Where are the three Shaktis? Where are the six cakras? Where are the twenty one macrocosms? Where are the seven underworlds?

Ishvara said-The upper Shakti is in the throat, the lower Shakti is in the anus. The middle Shakti is in the navel and the Shakti above these is without description.

The adhara cakra is in the anus, the SvadhishÖhana is near the penis. I have spoken previously of the distinctions of the cakras. Hail to that above the cakras!

The top half of the body is called Brahmaloka, below that is Patalaloka. The body resembles a tree with the roots at the top and the branches below.

Devi said-Parameshvara, Shiva, Shankara, Ishana. Speak to me! What are the ten vital breaths and what the ten doors?

Ishvara said-Prana is in the heart, apana is in the anus, samana is in the navel region and udana is situated in the throat.

Vyana is spread throughout the body, Naga moves upward, Kurma is situated in the tirthas.

Krikara is in crying, Devadatta is in yawning, Dhananjaya is in singing and roaring.

According to the yogis, these vital breaths have no support. The nine doors are clear and the tenth door is the mind, it is said.

Devi said-Speak to me of the characteristics of the nadis spread throughout the body. Tell me of the ten nadis arising from Kundalini Shakti.

Ishvara said-Ida, Pingala and Sushumna go upwards. Gandhari, Hastijihva and Prasava move below.

Alambusha and Yasha are situated to the right. Kuhu and Shankhini are situated to the left.

Various nadis, amounting to seventy two thousand, arise from these ten and exist in the body, it is said.

Only a yogi with knowledge of the nadis is a true yogi. Devi, nadi knowledge gives siddhi for yogis.

Devi said-Bhutanatha, Mahadeva, Parameshvara, speak to me! What are the three devas? What are the three bhavas? What are the three gunas?

Ishvara said-The Rajobhava is Brahma. The Sattvabhava is Hari, The Krodhabhava is Rudra. These three devas are the gunas.

These three devas, Brahma-Vishnu-Maheshvara, are one. Whosoever thinks of them as different in nature is never liberated.

Brahma is like semen, Hari is like vital breath and Rudra is like mind. The three devas are the gunas.

Brahma is compassionate, Hari is pure and Rudra is fiery. The three devas are the gunas.

With one meaning, they are the same thing, supreme Brahma, the cosmos, all which moves and does not move. Whosoever thinks of them as different in nature is never liberated.

I am creation. I am time. I am Brahma. I am Hari. I am Rudra. I am Shunya. I am everywhere. I am without distinction.

I am the soul of all, O Devi. I am free from desire, I am space. I am my pure own nature, without stain, there is no doubt of it.

He who has controlled his senses and courageous is a Brahmacari and a true pandit, speaking truth, devoted, generous, steady of mind.

The path of Brahma is the root of tapas. Compassion is the root of duty. Therefore, by every means, rest in the path of compassionate duty.

Devi said-Yogeshvara, Jagannatha dear to Uma as life itself, speak of veda, twilight worship, tapas, meditation, and sacrifice.

Isvara said-One thousand ashvamedha and one hundred vajapeya sacrifices are not equal to even a sixteenth fraction of the greatness of Brahma Jnana.

Travelling to the sacred tirthas and bathing there does not amount to a sixteenth fraction of the greatness of Brahma Jnana.

No friend, no son, no father and no relatives can ever equal the guru, the revealer of the supreme thing.

Learning, tirthas, and gods and goddesses (devatas) can never equal the guru, the revealer of the supreme thing.

A disciple who has received a single syllable (of a mantra) from a guru can never find anything in the world to repay that debt.

The Brahma Jnana is secret knowledge and should not be revealed to anyone but a devoted soul.

The wise should reject mantra, puja, tapas, dhyana, homa, japa, animal sacrifice, nyasa, and all acts.

Associating with others has many defects, not associating with others has many qualities. Therefore, by every effort, the wise should reject the company of others.

The letter a is sattvik, the letter u is rajasik and the letter m is tamasik, these three are Prakriti itself.

The indestructible is Prakriti, Ishvara is Prakriti and is itsef indestructible. From Ishvara came Prakriti which is triple in nature.

Prakriti is Shakti, the Maya, the cause of creation and destruction, the Avidya, the Mohini, whose form is sound.

A is the Rig Veda, the letter u is the Yajurveda and the letter m is the Samaveda.

The Om should be known as the three nadas. The letter a is Bhuloka, the letter u is Bhuva and the letter m is Svarloka, it is said. In these three letters the Atma manifests.

The letter A is earth and yellow in colour, the letter u is like lightning and space, and the letter m is white and heavenly. For sure the single Om pervades everything and is Brahma itself. Those seated in a steady position, who meditate daily and are free of worries and imaginings, are yogis and not otherwise, Shiva said.

Whoever hears this or reads this everyday becomes pure and free of all problems and resides in the Shiva heaven.

Kaulavalinirnaya Tantra

This is Sir John Woodroffe's (Arthur Avalon) introduction to a Sanskrit edition of the *Kaulavalinirnaya* in Sanskrit which is now out of print and out of copyright. Because it covers many topics relating to the Kaula tradition of Tantra, it merits wider availability. The text, to the best of my knowledge, is not available in an English translation.

Kankalamalini Tantra

The word *kankalamalini* means garlanded in bones, or skeletons, a constant theme of texts related to the goddess Kali. This work, therefore, falls into the Kalikula category of tantrik texts. The edition used for this abstract was published by Kalyana Mandir Prakashan in 2033 Samvat (1973 c.e.).

The *Kankamalinitantra* is a relatively short work of only five patalas (chapters). Like many tantras, each is of uneven length. Chapter five is probably longer than the previous four chapters put together. In the colophon at the end of each chapter, the Tantra is ascribed to the Dakshinamnaya, or southern tantrik current. Mantras are given below using the itrans format to preserve their correct form and pronunciation.

Chapter One

The first chapter opens with Bhairavi asking Bhairava to tell her about the letters of the alphabet. He says that the letters A to Ksha form the absolute as sound (*Shabdabrahma*), and then proceeds to enumerate the female shaktis associated with these letters. Bhairava says that the letters A to Ksha consist of Shiva and Shakti, and without knowledge of their true meaning, it is not possible to be successful practising the Vama path. The letters are made up of the three gunas. This chapter, shorter than the others, then discusses the major bija or seed mantras, including OM, shriim (Lakshmi bija), kriim, kliim, hriim, hum, huum, hraum, aim, kraum, svaahaa, driim and duum, priim, Tham Tham and sphriim.

Chapter Two

Bhairavi addresses Bhairava as Nilakantha (the blue throated) Mahadeva and asks him to explain to her the meaning of yoni mudra and the three tattvas. He says that yoni mudra is very secret and should not be revealed. It is the very form of the absolute, representing the chaitanya or consciousness of mantra and bestows liberation. By grace of the yoni mudra, Bhairava says he was able to conquer death. Semen, blood and their conjunction are the temple of Manmatha (the god of love). The yoni bija mantra should be recited 108 times and the yoni itself is the true form of the supreme absolute. The yoni should be encircled with three threads, which are the ida, the pingala and the sushumna. The yoni of Devi is the primordial (Adya) form of Prakriti (nature). It is Kundalini and Mahakundalini, says Shiva.

Bhairava then speaks of the nadis or channels of energy in the body. There are 3.5 koti (a koti is 10 million) nadis, but the three nadis mentioned above are the chief, and represent the moon, the sun and fire. Bhairava then goes on to describe the six well known chakras through which runs the thread of the sushumna or central nadi in the spine. Details are given of the Dakinis, the gods and goddesses, and the bijas of each of these chakras, with very similar details to those published by Sir John Woodroffe in *The Serpent Power*. Above the Ajna chakra, says Bhairava, is a lotus of 1,000 petals, which is the place of the seventeenth kala.

Kundali Shakti is the form of mantra, dwelling in the muladhara chakra and rising through the Chitrini to the Brahmanda or 1,000 petal lotus, is the rosary of letters, says Bhairava.

There then follows a Yoni kavacha which is of some interest. Ishvara says that by holding it and reading it, it causes all shaktis to give boons.

The rishi of the kavacha is Sadashiva, the metre for it to be pronounced in is Gayatri, the devata is the Eternal Yoni form, while it gives the four aims of mankind.

OM mam maam mim miim mum muum mem maim mom maum mah mama shiro raxantu svaahaa.

OM mam maam mim miim mum muum mem maim mom maum mah OM maam OM aakuutaam mama raxantu svaahaa mam maam.

OM mam maam mim miim mum muum mem maim mom maum mah hridayaadi daxa bahum raxantu.

OM mam maam mim miim mum muum mem maim mom maum mah hridayaadi vaama bahum raxantu.

OM mam maam mim miim mum muum mem maim mom maum mah daxa paadam raxantu mama.

OM mam maam mim miim mum muum mem maim mom maum mah vaama paadam raxantu mama sadaa svaahaa svaahaa.

OM mam maam mim miim mum muum mem maim mom maum mah hridaadisu naasaam raxantu svaahaa.

OM mam maam mim miim mum muum mem maim

mom maum mah upastham raxantu mama sadaa svaahaa.

OM mam maam mim miim mum muum mem maim mom maum mah idam hi yoni kavacham rahasyam paramaadhbhutam.

The kavacha should be recited in the Muladhara, before the eternal yoni. It gives equality with the sun and the moon, and through the grace of Devi causes success in the yoni mudra. The text says it should be recited with one's own woman or with another woman, following which there should be intercourse. This is an example of tantrik code. The "other woman", according to tantrik insiders, is one's wife or woman, while one's own woman here refers to the Devi within. From this point of view, sex with one's own woman is adultery.

On the other hand, this Tantra may well be speaking literally. The kavacha, continues Ishvara, should be written on bhurja leaf (birch) and written with svayambhu flowers (menstrual blood), and semen, and with other scents such as gorochana. It should be placed inside a gold ball and worn on the body. Reciting the kavacha 108 times gives success in whatever is desired. The chapter closes with the mantra namo yonyai namo yonyai kundalinyai namo namah.

Chapter Three

This chapter returns to comparatively more sedate matters, including guru puja, the guru mantra and a guru gita (song).

Ishvara says the two syllables of the word guru represent that which is without qualities and the supreme absolute, respectively. This mantra, he says, is the mahamantra, and should be concealed.

He follows by giving a dhyana of the guru, situated in the 1,000 petal lotus. He is seated in the virasana, with his two hands showing the mudras dispelling fear and giving boons. On his left thigh sits his shakti, whose face shows compassion. She wears red clothes and jewels.

The guru and his shakti should be worshipped with mental offerings (upachara). After reciting the mantra of the shakti, the kavacha of the guru should then be read. This kavacha differs from that given in the *Matrikabhedatantra*. Wearing the kavacha on different parts of the body washes away demerit in the same way as the Ganges river washes it away. The chapter closes with a brief song (gita) extolling the virtues of the guru.

Chapter Four

This chapter centres around the worship of Mahakali. Parvati asks Shiva to give the Kali mantra, and describe her puja.

Ishvara (Shiva) says that the mantra of Mahakali bestows every type of success. All the gods and rishis achieved what they did through her worship. It gives both liberation and enjoyment and bestows liberation through enjoyment, when heard from the mouth of the guru.

Shiva gives the mantras of Mahakali and says her one syllable mantra is the gives siddhi in the Kali age. He then gives a three syllable mantra of Dakshina Kalika, followed by other three syllable mantras producing different results.

Conventional rules in the worship of Mahakali are suspended, says Shiva. There is no rule as to time, as to the woman who is the shakti, or to defects of the mantra. Similarly, one need not pay attention to bodily defects. The sadhana may be done during the day or at night, and the recitation (japa) of the mantra may be done anywhere.

By pleasure one gains liberation, Shiva says. He says: "This is true, true, true and again true, I say." Puja is of three types, he says: daily, every so often, and according to desire (kamya). Here he says he will speak of the daily puja of Mahakali. Bhairava is the rishi, Ushnik is the metre, and the devata is Mahakali, who gives the four aims of mankind. Without fivefold purification, any puja undertaken is black magic (abhichara).

Those five purifications are of the atma, bath, the materials, the mantra and the devata. Following the placing of the materials, one should bow to the gurus on the left and to Ganapati on the right, and should then perform bhutashuddhi, the purification of the elements. (This is a meditation in which the different elements within a sadhaka are purified.

Different nyasas are then performed, and the text follows with a meditation image of Mahakali. One should meditate on Adya Mahakali as being in a celestial spot, on the central peak of the Himalaya range, under a jewelled pavilion which is the great pitha, her lotus feet served by Narada and the best of saints, worshipped by Bhairava. She is the colour of sapphire, with two large high breasts, wears variegated colour clothes, and has four arms and three eyes.

The text then follows with a description of her inner meditation, where she has limbs the colour of thunderclouds, dishevelled clothing, three eyes and is seated on Shiva's corpse. She is ornamented with a chain of skulls. In her left upper hand she holds a man's severed head, and with the lower hand holds a cleaver. She has dishevelled hair.

Shiva then gives a further dhyana of Mahakali, where she has a fierce, fanged mouth, is completely naked, and has three eyes. She sits in virasana on Mahakala and makes a terrifying noise, wears a garland of skulls (mundamala) and has streams of blood pouring over her full breasts. She sways backwards and forwards, as if intoxicated.

In her left hands she holds a cleaver and a severed head, and in her right shows the mudras giving boons and dispelling fear. She has a terrifying face and her tongue rolls wildly. She has earrings made up of a bird's wing and an arrow. She is served by terrifying, roaring jackals in the cremation ground and by Bhairavas making fearful laughing noises, and who dance over men's skeletons, making their victory cries.

The text then follows with a description of Kali's fifteen attendants, the Kalinityas. This leads up to the left-hand worship with the panchatattva or panchamakara. Shiva says whoever does Kula puja without wine or flesh loses the merit of 1,000 good incarnations. "Without wine, there is no mantra, there is no mantra except with wine," Shiva says. After performing the rite of the five makaras, one should bow again to Mahakali before doing the dismissal and cleansing rites.

Chapter Five

This chapter is concerned with purashcharana, the rites to be performed by an initiated tantrika to make a mantra successful. Elsewhere on this chapter, we have described the elaborate rules, stretching over several days, which a sadhaka has to undertake. But the *Kankalamalini*, in a similar matter to the *Devirahasya* and the *Brihadnilatantra* seems to suspend these rules.

Parvati is told by Ishvara that in the Kali Yuga, folk are short lived and unable to perform rites in the way they were able to do in previous times. He says that for this worship there is no bad time, no special day or night, no need to do the puja on "great nights", such as the eighth or fourteenth of a dark fortnight, nor is there a necessity for worship at the twilights (sandhya).

There are no rules about directions, places, recitation of mantra, time to do the worship. "Here, svecchacharya (doing the rite according to will) is the rule for the mahamantra in sadhana," Ishvara says.

Performing worship in the Kali Yuga in this fashion brings siddhi in six months, according to the text. Shiva says: "Devi, in the Kali Yuga, there are no tirthas (bathing spots), no vows to undertake, no homa, no bath, and no twilight worship (sandhya). "Those rites belong to the previous eras of the Saya, Dvapara and Treta Yugas, he says.

However, purashcharana is still necessary, he adds, and proceeds to give the rite suitable for tantriks during the Kali Yuga. There then follows a lengthy rite which includes the giving of substances including ghee, milk, and sugar, and the recitation of many mantras, the performance of many nyasas. The importance of the rudraksha rosary is stressed at great length. The sadhaka should smear himself with ash, and put three lines on his forehead as well as a tilak.

Rules are given about the use of the Gayatri mantra, and towards the end of chapter five, there is a lengthy discourse on the Devis of the bodily dhatus, such as Dakini, Lakini, Rakini and so forth, along with their bija mantras and their various meditation images. These Devis are situated in the different chakras.

Kulananda Tantra

Like the Akulavira Tantra, this text is also ascribed to Matsyendranath, described in the colophon as Matsyendra Pada. This Tantra was also published in Bagchi's original *Kaulajnananirnaya of the School of Matsyendranath*, but omitted from the edition published by Prachya Prakashan in 1986. Bagchi found the text in the Darbar Library, No.135 and says it is transcribed in Newari script.

This text is, however, briefer than the *Akulavira* at only 60 verses, with a few lacunae and omissions marked with asterisks. We suspect the worst here-that these are verses with sexual content and which Bagchi has chosen to leave out. The Tantra follows a more familiar tantrik pattern, with information about the chakras, the granthis, and piercing the chakras. Opening with the salutation OM namo bhairavaaya, the first four verses describe a familiar scene on Kailash mountain, where Devadeva, the world guru, is questioned by Uma Devi. However, she wants to know about practices and methods relating to the destruction of old age and death, of samarasa, and of techniques and results of the siddhas, many of which are described in the *Kaulajnananirnaya*. The text is given below in itrans format.

In verses seven to 16, Bhairava describes the place of the supreme where a sadhaka should meditate, as well as the heart lotus and the granthis, or knots. Devi asks him about piercing (bheda) these, as he has not spoken of them before.

In reply, Bhairava (vv 18-23), begins by describing the Brahmasthana lotus, which he says has 64 petals, and follows with descriptions of other lotuses including the brow lotus between the two eyes, which appears, although the reading is unclear, to conquer time, as well as giving the ability to see and hear things going on at a distance.

In verse 24, Devi asks about removing wrinkles and decay. Bhairava says (vv 25-34) that while this knowledge is hard to get even for the gods, he will speak of it. A 32 petal lotus exists in the head, and there one should meditate on nectar, made up of the 16 kalas of the Moon, which one should cause to fall to the heart, and then towards the navel chakras. Meditating on the nectar as black in colour, it removes wrinkles and the like within the space of a month, he declares.

Another method, he says, involves meditating on the place of 64 petals (described above), the place which is the abode of all the gods. Meditating here destroys the effect of poison, fever and disease. Another method, says Bhairava, seems to be a meditation on junction places, which appear to be situated in the head, the place of Urdhva Shakti, as bright as millions upon millions of fires.

The goddess asks, in verse 35, about methods called dhunanam and kampanam, practices which are often referred to in the *Kaulajnananirnaya*. Bhairava replies at some length on these and other topics (vv 36-54). These siddhis seem to be achieved by meditating on the Mahavaha Nadi, described as situated in the kuchelya, which may be the same as the bankhnala. There is a lacuna in the next verse which refers to the heart chakra and the Tivrachakra. Meditating here bestows dhunanam and kampanam siddhis. In verse 38, Bhairava speaks of khechara. Shakti has the appearance of a coiled serpent (kundalya). When she reaches the head, it produces khechara siddha. Bhairava then describes a process involving five svaras or syllables, which a sadhaka is to meditate on concentratedly. Another meditation on the vayus is described.

The state of samarasa is described towards the close of this brief text. Here one is to meditate on Devi as abhava, free of all distinctions. The methods outlined in the Tantra, he concludes not only give the ability to see and hear at a distance, but allows yogis to enter other's bodies.

kulaanandatantramh

OM namo bhairavaaya

kailaasashikharaasiinam devadeva jagadgurumh.

pariprichChatyumaadevii ekaante gyaanamuttamamh..

bhedanimuttamabhedam yathaa dehavyavasthitamh.

kathayasva puraabhedam kulaanandeshu choottamamh..
sthaanaantaravisheshana vigyaanam kathayasva me.
sadyah pratyakaarakam yathaa dehe vyavasthitamh..
paashastobhancha bedhanancha dhuunanam kampanam tatha.
khecharam samarasanchaiva baliipalitanaashanamh..
sarvvam tattu sureshvara kathayasva mama prabho..
bhairava uvaacha
shrinu devi pravaxyaami purabhedam samuttamamh.
etath kaulikam gyaanam kulaanande chaashtottamah..
brahmasthaane yath kamalam chatuhshashthi-dalaanvitamh.
tatraiva manasaa rodhya laxayeddiipashikhaanh vratii..
pashyati sarvvadehe tu divyadrishtirvaraanane.
tena laxitamaatrena jaayate siddhiruttamaa..
tasyopari naabhimnadavadaa vichaxane.
tathaa choparitishthati devo bhairavo jagathprabhuh..
tasyaagre manasaaropya dridhabhuutavichaxanaah.
jaayate dhvanibhirjaatim tatra marudvisheshatah..
etadh vigyaanamaatrena aavesham jurute dhruvamh.
jaayate paramam sthaanam naatra kaaryyavicha-aranaath..
athaanyam paramam devi sadyah pratyaakaarakamh.
tam pravaxyaami he devi shrinushvaayatalochane..
angulyadvitiyam devi yath puurvvam kathitam mayaa.
tatraiva sthaapayedvittam kathitam jaayate dhruvamh..
atiitaanaagatanchaiva varttamaanam tathaiva cha.
tatraivothpaadayeth srishtim tivrajyotihsamakritih..
hridisamsthakamalam hitvaa muurdhni yaanh prapuurayeth.
paashastobham karotyevam yadi vishramate manah..
athaanyam param devi divyadrishti pravarttate.

jvalajvaalanamadhyastham dvijagranthi niriixayeth..
tatraiva manasaa rodhya chuulikaagrantu maanayeth punah.
srishtim pravarttante devi naatra kaaryyavicha-aranaath..
devyuvaacha
tadbedhancha shuddhasaarancha hitam mayaa na prakaashitamh.
kathaysva visheshana yathaa jaanaami tathxanaath..
bhairava uvaacha
shrinu devi pravaxyaami bedhasaaram yathaakra-mamh.
brahmasthaane padmam chatuhshashtidalaanvitamh..
tatraivothpaadayeth srishtih dandaakaaraagnitejasaa.
bhedayitvaa mahaagranthim dvichakrancha tathaiva cha..
muurdhnim kamalasamsthaanam jvalanaakaaram vichintayeth.
chalantam bhraamayettena niyatam tu mahitale..
bedhaye girivrixaani kim punah xudrajantavah.
athaanyam paramam devi vigyaanam bhadrakam shrinu..
bhruvoshchaxurmadhyastham cha kaalaavunaa vichintayeth.
yaavadvishramate nityam bhramate chakravath shirah..
duraashravanavigyaanam bhuutadetasya chintayeth.
vaamastrotam tu yaavath tivratejasamaprabhamh..
devyuvaacha
baliipalitaxayam deva tvayaa khyaatam mamaagratah.
etanme samshayo deva kathayasva mama prabho..
bhairava uvaacha
shrinu devi pravaxyaami yath suraanaamapi durllabhamh
dvaatrimshatidalam padmam tishthati tanmuurdhni madhyatah..

tatraiva bhaavayedamritam kalaashodasha-samanvitamh.
hrichchakram paatayedvaaram yaavadaaana-abjimandalamh..
chintayeth kriShnavarnancha bhedam tu sakalam purah.
palitaastambhayeddevi maasamekena suvrate..
athaanyam paramam devi guhyasamshayasthi-tikaarakamh.
yatra chatushpathasthaanam tatra sarvvaasura-alayamh..
tatraiva sthaapayechchitam divyabaalapravarttata
ashrutaam vadate baalaamdivya varaanane..
vishaprahaaram kurute jvaravyaadhim vinaashayeth.
tatah parataram sthaanam tatra sarvve pravarttate..
athaanyam paramam devi duraabedham vadaamya-hamh.
jvalajjvalanasamdhyastham tatraiva laxayeddevi..
tam tritakotisamaprabhaa uurdhvam shakti nipaataanyatah
bedhayedvichaxanah bedhayitvaa tu tam laxayeth..
jvalanaakaashthaprabhaam urdhvashaktim chakrayitva achalam bhraamayeth punah
bedhayeth saa manahsahasraani samakaani tu kaa kathaa..
evam gyaatvaa viiraarohe vichareta yathaasukhamh.
devaasurmanujaanaam durlabho bhavati saadhakah..
devyuvaacha
dhuunanam kampanam deva yathaakhyaatam mamaagratah.
tatashchaapi samshayo dėva kathayasva mama prabho..
bhairava uvaacha
shrinu devi pravaxyaami dhuunanam kampanam sthitamh.

yaa saa mahaavahaa naadii kuchelyaakaara-samsthitaa..

tatraivothpaadayeth drishtirddandaakaarasutejasaa.
hrichchakra tivrashakti suvigrahaa..
dhuunanam kampanam chaiva naatra kaaryyavicha-aranaath.
athaanyam paramam devi khecharatvam shrinu priye..
yaa saa mahaavahaa naakundalyaakaarasamsthitaa.
padadvandvagataa saa tu muurdhnistham kamalam punah..
yaa viirataye tajjvalanaaakaaram subhaavayedh budhah.
sarvvashaktito madhyata khecharatva bhavedveh navaahamabhritam dadaitah..
athaanyam sampravaxyaami mahaavyaaptim varaanane.
yaavadh. sthaaneshu khyaataa tasya mayaagratah..
tathaiva manasaaropyam naasaagre pavanamaanena.
manasam tattva bedhayitvaa nishkalam yojayedh budhah..
mahaavyaaptirbhaveddevi tavaahamanritam vadeth.
athaanya sampravaxyaami yena siddhirbhaveth dhruvamh..
vargaatiitasya dvitiiyamanuloma na suvrate.
jvalanaaruudham kritvaa chaturtham svarasamyutah..
bindumastakasambhinnamaadau yojayedh budhah.
vargaatiitam punardevi shashtasvaravibhuushitah..
sa shikhe kharabindusamyuktam yogayeddevi tath punah.
yakaarasya panchamam grihyam tam tritiiyasvara-samanvitamh..
svaram dvitiiyasamyuktam kaarayeth punah priyah.
vargasya prathamam biijam srishtaraadyato yojayeth..
etaddevi samaakhyaatam kulavidyaasamanvitamh.
nityam tu yojayeddevi ekachittastu panditah..

tasya sarvvaa bhaveth siddhih sarvvagyaneshu sundari.
na khedayaami kasya chittam shishye.apyuktvaa varaanane..

athaanyam paramam devi kaamena vihvalaarakamh.
bhasya madhyastha bhagamadho hutaashanamh..
preritah pavanah shaktyaa jvalanaakaaratejasaa.
visantu chintayeddevi srishtikamalantu yaavath..
shravanam chintayedamritam tejomaargena yogini.
bhavanti vihvalaa naaryyah xanamekena suvrate..
athaanyam paramam devi phalapushpaakathanam cha.
shrinuyaath parmaadevi aghoravikramaa..
aakarshayedh bugho vaayum pindastham ruupanamuchyate.
chaxushaa yojayeddevi phalapushpantu chintayeth.
tejasaarkarshayeth tu chittam susaryyatah..
devyuvaacha
sarvvatra tu mayaa gyaanam bhaashitam deva tath puraa.
adyaapi samshayo devi kathayasva mama prabho.
samgamaanmoxahetukam
bhairava uvaacha
shrinu devi pravaxyaami tath surairapi durlabhamh
aasanam tu sthitam krtvaa vishva anivrittah..
abhaavam bhaavayeddevi sarvvabhaavavivarjjitamh.
chittam tatra sthitam kritvaa khamadhye viniyojayeth..
bhaavayeth samarasam devi prabhubhrityam vichaxanah.
etadvigyaanamaatrena giyate tatra madhyatah..
duuraashravanavigyaanam bedhastobham madhyatah.
aavesham darshanam duuraath kampastobham tathaiva cha..
parakaayapraveshena sampravartteta yoginah.
tadabhyaasena suvrate..
iti matsyendra paadaavataaritam kulaanandam samaaptamiti.. shubhamastu sarvvajagataamh.

Kulachudamani Tantra

The Kulachudamani-Tantra or "crest-jewel" of the Kulachara division of Tantrika Sadhakas is included in the list of revealed works, which according to the Vamakeshvara Tantra, are considered to be the chief amongst those which deal with the worship of Shakti. It is accordingly found frequently referred to as an authority in many compilations though the Kulachudamani itself (II, 8) refers us, for all technical terms, to the Bhairavi Tantra, which is, however now known chiefly from quotations made from it.

Like all original works on Tantrika worship, the Kulachudamani is cast in the form of a dialogue—the Shastra being revealed by the Devi in Her form as Bhairavi, in answer to questions put to Her by Shiva in His form as Bhairava. For this reason the book is included in the class which goes by the name of Nigama as opposed to Agama, in which the Shastra is revealed by Shiva Himself. The form in which a Shastra presented whether as the Revelation of Shiva or Shakti is mere Lila. Since Shiva and Shakti are one and the same and it is Shiva who reveals, Shiva is the revealer of the Shastra in all cases, though in some He figures as Shishya and in others as Guru. The Svachchanda Tantra puts this clearly in the following verse:

Guru shishyapade sthitva svayameva mahesvarah

Prashnottara-padair vakyaistantrang samavatarayat

The Tantra according to this verse, was originally revealed by Maheshvara (Shiva) who Himself stood for that purpose in the position of the Guru as well as that of the Shishya. This is also stated in the last chapter (VII, 79) of the Kulachudamani where the Devi addressing Her lord says:-

Gurustvam sarvatantranam.

Kulachara has been called a secret doctrine and practice. The Bhairavi in discoursing of it in the Kulachudamani says at the outset (I,31) that it had not been told to Vishnu nor to Brahma nor to Ganapa.

Kaula knowledge, says the text (II, 25) must not be divulged to atheists, fools, Pashus or to persons of the twice-born caste.

The secret teaching appears to have been transmitted for a long time "from mouth to mouth" (Vaktrat vaktrantaram) and even when it came in part to be reduced into writing, sufficient precaution was taken to conceal it from the uninitiated under technical terms, the import of which could only be learnt from the Guru. The general features may, however, be summed up as follows.

Although the word Kula in ordinary parlance means a family or clan, its technical sense has been defined by the Tararahasyavrittika to be "Kulam=matri-mana-meyam". The term thus combines the meaning of the three other words which are further explained to mean Jiva (Mata), Jnana (Manam) and the manifold universe or Vishva (Meyam). The gist (Sangkalitartha) therefore is said to be Shakti. As Shakti is Kula so Shiva (as distinguished from Shakti) is spoken of as Akula. Kulachara is one of the seven Acharas enumerated by the Kularnava, one of the leading Tantras of the division of Sadhakas of this school called Kaulas. According to the last named Tantra, it occupies the highest rank. "The Vaidikachara" it says, "is no doubt higher than all, but Vaishnavachara is higher than Vaidikachara; Shaivachara is higher than Vaishnavachara; Dakshinachara is higher than Shaivachara; Vamachara is higher than Dakshinachara; Siddhantachara is higher than the last and Kulachara is superior to all."

As Kulachara is thus said to be the highest of the Acharas, only those Sadhakas are qualified therefor who in this or another birth have graduated in the preceding Acharas which are regarded as stepping stones to it. Such a Sadhaka is called Kaulika or Kulina. Being the final stage of Sadhana this Achara knows no distinction of race, colour, caste, or sect. But the esoteric character of its doctrine and practice is such that it was never meant for the ordinary man of the world.

On the contrary, the difficulties of its true practice are said to be such that according to the doctrine "it is easier to walk on a drawn sword," than to be a true Kaula. It is expressly stated (I, 42) that the Adhikari must be a Kulina, that is one who is capable of realizing that every person, thing and act is a manifestation of the Mother or Shakti (Striyamancha jagatsarvam). An essential feature of this Achara is the attainment of the knowledge that the Mother who is worshipped under different forms as Tripura, Kalika and so forth with differing rituals is She from whom all creation proceeds and who is all in all.

This is very aptly set forth in the text (I, 24) which says "Oh All-knowing One, if Thou knowest Me then of what use are the Amnayas (revealed teachings) and Yajanam (sacrifices: ritual). If Thou knowest Me not, then of what use are Amnaya and Yajanam."

Yadi mang viddhi sarvajna kva chamnayah kva yajanam,
Na viddhi mang chet sarvajna kva chamnayah kva yajanam.

This teaching has found its way into popular Bengali songs which say.

"Tell me what will japa, tapa, yoga and yaga do for a man in whom Kulakundalini awakens and for the man in whom She awakens not." Supremacy is claimed for Kulachara on the grounds that it is the final stage of Sadhana in which Knowledge is realised to be superior to ritual. Kuladharma is accordingly said to weigh more than all Yajnas and Vratas put together in the scale against it though such rituals are necessary in the preliminary Acharas which qualify for the last.

As Jnana alone secures liberation the Kularnava Tantra affirms that without Kuladharma liberation is not possible. With the question whether this claim is well founded I am not concerned but with the statement of the historical facts. As being the Achara which is claimed to be at the entry of liberation it is regarded by Kaulas as supreme and the end for those which precede it. Other schools take a different view of the Kaula claims.

Chapter I

The book opens with an enumeration of the Kula-sundaris or Devis who are said to be innumerable under the names of Tripura, Kalika, Vagishvari, Sukula, Kula, Matanggini, Purna,

Vimala, Chandanayika, Ekajata, Durga and others. Several doctrines also such as Vaishnava, Ganapatya and others are mentioned.

The names of a number of Tantras belonging to the sixty-four are next given. A complete list has been quoted in the footnotes (pages 2-3) from the Vamakeshvara-Tantra, according to which each of the eight Bhairavas has a Tantra of his own, all of which are collectively known as the Bhairavashtakam. Similary the Tantras relating to the seven Matrikas the Shivadutis are collectively called Bahurupashtakam. The Yamalas are eight in number. These three classes give us 24 works, while the rest are those named in the list.

The 64 Tantras given in the Vamakeshvara are as follows:

1. Mahamaya
2. Shambhara
3. Yoginijala-shambhara
4. Tattva-shambhara

5-12. Bhairavashtaka: (a) Asitangga; (b) Ruru; (c) Chanda; (d) Krodha; (e) Unmatta; (f) Kapali; (g) Bhishana; (h) Sanghara.

13-20. Bahurupashtaka-the eight Tantras of the seven Matrikas and Shivadutis

21-28. Yamalashtaka: (a) Brahmayamala; (b) Vishnuyamala; (c) Rudrayamala; (d) Lakshmiyamala; (e) Umayamala; (f) Skandayamala (Bhaskara substitutes Jayadrathayamala); (g) Ganeshayamala; (h) Grahayamala

29. Mahochchushya
30. Vatula
31. Vatullotara
32. Hridbheda
33. Tantrabheda
34. Guhyatantra
35. Kamika
36. Kalavada

37. Kalasara
38. Kubjikamata
39. Tantrottara
40. Vinatantra
41. Trodala
42. Trodalottara
43. Panchamrita
44. Rupabheda
45. Bhutoddamara
46. Kulasara
47. Kuloddhisha
48. Kulachudamani
49. Sarvajnottara
50. Mahakalimata
51. Mahalakshmimata
52. Siddhayogeshvarimata
53. Kurupikamata
54. Devarupikamata
55. Sarvaviramata
56. Vimalamata
57. (a) Purvamnaya; (b) Pashchimamnaya; (c) Dakshinamnaya; (d) Uttaramnaya
58. Niruttara
59. Vaisheshika
60. Jnanarnava
61. Virabali
62. Arunesha
63. Mohinisha
64. Vishuddesvara

It has not been found possible to identify with certainty all items in the list given in this work with that of the Vamakeshvara and in some respects the list differs. If, however, we take Mayottara, Kalapaka or Kalapada, Sarvajnatmaka and Vishudeshvara which occur in the Kulachudamani to refer to items 1, 36, 48 and 64 respectively of the Vamakeshvara list, then the two lists correspond except as to the 9 items marked with an asterisk. In the place of those last the present work appears to give the names of the following Tantras:- Mahasarasvata, Tantrajnana, Vasuki, Mahasammohana, Mahasukshma, Vahana, Vahanottara, Matribheda, Vishvatmaka Shivavali. If, however, we thus count them we get to (sic) Tantras or one too many. Possibly Mahasukshma may be part of the title of the Vahana Tantra which succeeds it, in which case it may be eliminated.

The Bhairava then says that He knows all these Kulasundaris, doctrines and Tantras but has nevertheless not attained bliss (Ananda). He asks the Bhairavi why this is so?

In reply, the Bhairavi first gives some general philosophical instruction in eleven verses (I, 16-26) to the Bhairava whom She addresses as the most Supreme Kula the ocean of Tantrika Kula knowledge (Tantrajnanakularnava) which, since He apparently seeks instruction, has for the moment been obscured by her Maya. This portion may be divided into three sections. The first (vv 16-17) refers to that primordial state when She as Prakriti was hidden in Chidananda (Ahang Prakritirupa chech chidanandaparayana). In this state there is neither creation, maintenance or destruction; neither brahma, Hari or Shambhu or other Devas, neither attachment, suffering nor liberation; neither piety, Theism or Atheism, Japa, Guru or Shishya.

The second state (vv 17-24) is that in which the Devi covering Herself with Her own Maya becomes desirous of creation (Unmukhi) and threefold. Then joyful in the mad delight which comes of Her union with the supreme Akula She becomes Vikarini; that is the Vikaras or Tattvas arise in Mulaprakriti.

Mayayachchhadya chatmanang tridha bhutva yadonmukhi.

Parakularasonmadamodini cha vikarini.

At this second stage Brahma, Hari, Shambhu appear and with them the Worlds (Loka) and the Elements (Panchabhutani) of which they are composed. By the differentiation of Shiva and Shakti the Gunas commence to operate (Shivashakti-prabhedena gunotpattistu jayate). Brahma and the others are not distinct entities. They are all one and the same as parts of Her. The creation which is Matratmaka appears and then disappears in Pralaya.

In the third section (vv 24-26), the Devi teaches the great lesson that all scripture and ritual are unneeded where She is known; as they are unneeded where She is not known. For scriptural teaching is a means to an end:-knowledge of Her. It therefore has no use where She is known. If on the other hand religious disposition is wholly wanting these means alone will not evoke it, though they are not without their uses in educating a latent piety in the disciple. The Bhairavi then says "I manifest myself as woman (that is in female form or Shakti) which is my own Self and the very essence of creation (Narirupang samasthaya shrishtisaram madatmakam) in order to know Thee Bhairava, the Guru who are united with Me (Bhavayogastham)". She adds that even when all this is said Her Tattva is not known.

The Devi then speaks of the methods (Upaya) of attaining liberation which is the essence of all Tantras and is honoured by all Devatas. These means secure knowledge and awaken Tattvabodha. They destroy both merit and demerit and (v 29) give both enjoyment and liberation (Bhogamuktipradayakam). This doctrine is said to have been kept as a profound secret so that it had not been divulged even to Vishnu, Brahma nor Ganapa.

It should be concealed in the heart (Gopaniyantu hridaye). "This wonderful secret, my child, should be kept from Pashus" (v 40. Rahasyam adbhutang vatsa gopavyam pashushangkate). The Devi speaking of this doctrine thus addresses the Bhairava "Child (Vatsa) it strikes me with wonder and bewilders even the wise. It is replete with numerous and bewildering meanings and is the final resting place of all good disciples (Sachchhishya-paramaspadam). It is Sadachara according to all doctrines (Sarvavadisadachara) and is at the same time blamed or reprobated by all doctrines (Sarvavadivigarhita). It can be learnt only from a good teacher (Sadacharyyaparijnaptam). Follow it with care."

To begin with, the Devi speaks of the necessity for the acquisition of Jnanashuddhi, the purification of knowledge, and for this purpose She refers to the daily observances beginning with the morning rites. The Sadhaka should rise in themorning, make his Pranama to the Kula trees (Kulavriksha), and contemplate upon the Kula (Shakti) from the Muladhara to the Brahmarandhra and meditate on the Guru. The Kulavrikshas, according to the Kameshvara Tantra are Sleshmataka, Karanja, Nimba, Ashvattha, Kadamba, Vilva, Vata, Ashoka. The Tararahasyavrittika quoting the above verse from the Kameshvara Tantra adds that the above are those usually enumerated but that a ninth is added by some namely the Chincha. The printed Tantrasara, however, gives a list of ten trees, viz, the first seven mentioned together with Udumbara, Dhatri and Chincha. From which it would appear that whilst Shleshmataka, Karanja, Vilva, Asvattha, Kadamba, Nimba and Vata are generally recognised as Kula trees, Udumbara, Dhatri, Chincha and Ashoka are only exceptionally so. Then follows the mental worship of the eight Kulanathas, namely Prahladananda, Sanakananda, Kumarananda, Vashishtananda, Krodhananda, Sukhananda, Jnanananda and Bodhananda. Their Dhyana is given in two verses.

They are those whose eyes betray the bliss in their hearts which comes from the great Rasa (Maharasarasollasahridayanandalochanah); whose darkness (Tamas) has been cut and crushed by embracing Kula; the dispellers of fear who know the meaning of all the Kula tantras (I, 36-37). The Chapter closes with the instruction that the Guru fit to initiate a disciple in this system must be a Kulina and no other. The Kulina is Adhikari of all Vidyas and is competent to initiate in all Mantras (Dikshaprabhuh sa evatra sarvamantrasya naparah). The work of those who leave the Kulaguru is stated to be mere Abhichara.

Chapter II

This chapter begins with the ablution—rites (Snana) and states the Shastric rules which must be followed. The devotee, after ablution, is directed to wear two pieces of cloth (II, 12). This is strictly in accordance with the rules laid down by Yogi Yajnavalkya. A departure from this rule constitutes, according to Bhrigu, nakedness, which disqualifies for the performance of religious rites.

The devotee is next directed to worship Shakti with offerings of flowers, incense, perfumed betel and other desirable articles. Then follows the worship with Yantra and the contemplation of Oneness with the Mother. The worship of the Sadhaka's wife (Nijakantam II, 30) is described with the details of her initiation to be given where She has previously been uninitiated.

Chapter III

The rites prescribed for the night are disclosed in this Chapter. If the devotee worships a Parashakti, he should first initiate her if she happens to be uninitiated. The mantra for such initiation is referred to in three verses (III, 13-15).

The food to be offered to Shakti during the Purashcharana ceremony is enumerated at length (III, 22-26). The Shaktis worshipped are to be looked upon as eight Matrikas, and they should be named accordingly. The hymn to be recited in their worship is given, which shows that each of them is to be addressed as one of the aspects of the Mother Herself.

The hymn is called Karnejapa-stotra from the fact that each verse addressed to each of the Matrikas is whispered into her ear. In this worship the elder may bow down to the younger, one of superior caste to one of inferior caste, for the Shaktis selected for worship are each and all manifestations of the Mother. The following translation and accompanying notes are by the General Editor.

Om

Obeisance to Thee O Mother! O Devi!
The pure One Who art Brahma
Remove by Thy mercy all obstacles which beset me
And grant me liberation

Our great Lady! (Bestower of blessings!

Oh Devi! Who art the Supreme Bliss
Remove by Thy mercy all obstacles which beset me
And grant me liberation

Kaumari! beautiful Playmate of Kumara!
The sovereign Mistress of all Vidyas!
Remove by Thy mercy all obstacles which beset me
And grant me liberation

O Devi! Who borne by the son of Vinata
Art Vishu
Remove by Thy mercy all obstacles which beset me
And grant me liberation

Oh Devi! Bestower of blessings! Who art Varahi
By Whom the earth was lifted on Thy tusks
Remove by Thy mercy all obstacles which beset me
And grant me liberation

Oh Devi! Who art Shakra
Who art worshipped by Shakra and other Suras
Remove by Thy mercy all obstacles which beset me
And grant me liberation

Chamunda! besmeared with blood wearing a garland of severed heads
Destructress of fear!
Remove by Thy mercy all obstacles which beset me
And grant me liberation

Mahalakshmi! Mahamaya!
Destructress of anguish and sorrow!
Remove by Thy mercy all obstacles which beset me
And grant me liberation

Thou art O Devi! the Father and Mother of all
And art to us in the place of Father and Mother
One Thou art yet manifold in the form of the Universe
Obeisance, Oh Devi! to Thee.

With the recitation of this hymn ends the Puja of the Shaktis; the worship of at least one of whom is enjoined. Other rites are enumerated in the subsequent chapters.

Chapter IV

The devotee should we well versed in Vaishnavachara that is in Bhaktimarga before he can be permitted to adopt any of the special rites. He should be Udarachittah (large minded), Paraninda-sahishnuh (patient of ill said of him) and Upakararatah (one who does good to others). He is required to recite his Mula-mantra if he happens to come across a deserted temple, junction of four roads and such other places which are said to be fit for worship. He is directed also to bow unnnoticed to Mahakali if he sees certain birds and animals; Gridhra (vulture), Kshemangkari (Brahmani kite), Jambuki (she-jackal), Yamadutika (raven), Kurara (osprey), Shyena (hawk, eagle or falcon), a crow and a black cat. He is further directed to do circumambulation (Pradakshina) of a corpse and the cremation ground.

The Mantras to be recited on such occasions are noted. Obeisance must also be made when he sees a black flower, red cloth, a king, a prince, an elephant, a horse, a chariot, weapons of war, valorous man (Virapurusha), a buffalo, a Kaulika or an image of Mahishamardini. If he sees a jar of wine, fish or flesh, or a beautiful woman, or a Devi Bhairavi, he must bow and recite a special mantra. The Nilatantra gives a fuller list of objects to which obeisance is to be made in this way. Then comes the ritual prescribed for the worship of Kali in the cremation ground. The Dhyana is revealed in seven verses (IV, 39-45). As this Devi fulfils all Siddhis She is called Dakshinakali.

Chapter V

This Chapter deals with rites which are performed for the development of powers enabling the devotee to draw towards him any Deva, lower Spirits or human being he wishes. Vv 7-8 say that if anything is taken from the subject of the rite which belongs to him or he or she are ill-treated or deceived in any way the Sadhaka is fallen (Bhrashta) and dies. Harm also happens to his family from such magic (Abhichara). The rite consists of the worship of Dakshinakali. The Rishi of the Mantra is Bhairava and Chhanda is Ushnik. The first Vija is the supreme Shakti (Purvang vijang parashakti). The Angganyasa is directed to be performed with the Vija coupled with the six long vowels. In this rite the Brahmana Sadhaka is directed to substitute for wine (where is this is mentioned) honey in a vessel of copper (v 78) or he may perform the Kulapuja with Kula wine.

Chapter VI

This part is concerned with the method of acquiring powers (Vetalasiddhi) enabling the devotee to go anywhere he pleases. It consists in the worship of Yoganidra, Katyayani, Purneshi, Chandi, Kamakhya and Dikkaravasini. The special rite of Sadhana which goes by the name of Shavasadhana is described in this chapter (vv 19-28). The object of this Sadhana in this special instance is the acquisition of the power with which the Chapter deals.

Chapter VII

This, the last chapter describes the worship of Mahishamardini which, from the large number of stone and metal images discovered in various places, seems to have been very popular at one time. The worship of Mahishamardini appears to have undergone gradual changes. This is indicated by the Mantra as disclosed in the Kulachudamani. The Mantra has been revealed in the usual Tantrika garb by the following verse:

Trailokyavijabhutante sambodhanapadang tatah.

Srishtisangharakau varnau vidya mahishamardini.

This yields a mantra of nine syllables, namely.

Ong Mahishamardini Svaha.

But the text (VII, 5) ordains that if the Mantra and its Sadhana is disclosed at all, it may be disclosed to one who is extremely obedient to his Guru, but even then not with its Vija. Only eight syllables should be disclosed, thus reducing the mantra to Mahishamardini Svaha.

It is said that the Mantra of nine syllables should not be imparted but should in the Kali age be kept concealed, and that eight syllables alone should be disclosed with the Mantra,

Svaha, but never with the Mantra Namah. It would appear from the Sharadatilaka, a compilation by Lakshmana Deshika of the eleventh century AD that in his day the Mantra of eight syllables alone was known.

The Mantra of ten syllables is not mentioned even by the Kulachudamani. This may be taken to suggest that the worship of Mahishamardini is of great antiquity. Originally the mantra was of eight, nine and ten syllables. But in course of time (at the date when the Kulachudamani was reduced into writing) the Mantra of ten syllables had already fallen into disuse, while the Mantra of nine syllables was discontinued. In the eleventh century AD (at the dayė of the compilation of Sharadatilaka), the Mantra of eight syllables only was known.

Another important change is noticeable in the Rishi and Chhanda of this Mantra. The Rishi, according to Raghava's Commentary on the Sharadatilaka, is said to be Shakavatsa; the Chhanda according to it is Prakriti. The Commentator also notices that in his day according to some, the Rishi was also said to be Markandeya. But the Kulachudamani (VII, 11) distinctly says that the Rishi is Narada and the Chhanda is Gayatri. It is noticeable that the Rishi and Chhanda of the Mantra for the worship of Durga are Narada and Gayatri. May it be that the worship of Mahishamardini was gradually sought to be cast into the same form as the worship of Durga? This seems highly probably from another circumstance that according to the Sharadatilaka-tika of Raghava Bhatta, the Pithapuja should be performed as ordained for the worship of Durga.

The image of Mahishamardini is, however, different from that of Durga. Mahishamardini, according to Kulachudamani (VII, 13) has eight hands holding on the right side Chakra (discus), Khadga (sacrificial sword), Vana (arrow), Shula (trident), and on the left side Khadga, Charma (shield), Dhanu (bow) and Tarjani-mudra (*vide post*). The Devi is said to be of black colour, wearing yellow cloth, and is placed on the body of a black buffalo.

This Dhyana does not exactly correspond with the one which is noted in the Sharadatilaka. According to it, the Devi holds in Her hands Chakra, Shangkha (conch shell), Kripana (sword), Khetaka (club), Vana, Karmuka (bow), Shula and Tarjani-mudra. The Devi who is said to be of the colour of Garuda stone (emerald) and bedecked with the crescent moon is described as sitting on the head of the buffalo.

In the Hymn incorporated with the text of the Kulachudamani (VII, 33), the Devi is said to be black of colour, resembling crushed antimony and is described as holding Chakra, Dara (Shankha), Kartrika (small sword), Kheta (club), Vana, Dhanu, Trishula and Abhaya-mudra. So far as this Mudra is concerned, Raghava cites an authority to show that Tarjani-mudra is the same as the Abhaya-mudra. It appears from the Tantrasara that the Devi is to be worshipped now in this form.

The Kulachudamani gives no Shangkha, or Khetaka or Katrika. It has instead two Khadgas and Charma. The Sharadatilaka mentions no Charma or double sword. It introduces Shangkha and Khetaka. The Hymn makes a further departure by changing the Khadga into Katrika. This is exactly what appears to have been in vogue when the Tantrasara came to be compiled about 400 years ago. The Hymn, therefore, appears to be of a date later than this. While the images appear to have changed in this way, the mode of worship has remained pretty fairly the same. The details are given with a view to help the reader to follow the ritual.

The worship of Mahishamardini is in general performed in the usual Tantrika way. The text only notices the points of difference which constitute its special features. The most noticeable of these is the Angganyasa which usually embraces six Anggas. In the case of the worship of Mahishamardini the text (VII, 15-17) mentions only five Anggas. The Sharadatilaka (XI, 25) says that in this worship Nyasa is made only upon five Anggas, leaving out the Nyasa of the eyes. The Dhyana is given in verses 12-14. The Yantra is composed of a lotus of eight petals, in each of which (VII, 18) eight Devis are worshipped, described generally as Durga and others (Durgadya).

Their names are given in the Sharadatilaka (XI, 29) as Durga, Varavarnini, Arya, Kanakaprabha, Krittika, Abhayaprada, Kanya and Surupa. They are worshipped with the long vowels a, i, u, ri, lri, ai, au and ah. Thus:-Ang Durgayai namah; ing Varavarninyai namah; ung Aryayai namah; ring Kanakaprabhayai namah; Lring Krittikaryai namah; Aing Abhayapradayai namah; Aung Kanyayai namah; Ah Surupayai namah. The Sharadatilaka-tika of Raghava Bhatta says that in selecting the long vowels, ri, lri should be rejected as neuter vowels. The Tantrasara, however, gives the long vowels as a, i, u, ri, lri, ai, au, ah. The weapons are also be worshipped along with the consonants beginning with ya or in other words the consonants beginning with ya, that is ya, ra, la, va, sha, sha, sa and ha are selected.

The hymn to Mahishamardini incorporated in the Kulachudamani is recited by Bhairava. The text of this hymn appears to have grown defective in course of time. Reference had accordingly to be made not only to the printed edition but to Ms. copies of the Tantrasara in which it is quoted. One Ms. dated 1604 Shaka year found in the district of Mymensing by the travelling Pandit of the Varendra Research Society, was of great help in restoring the correct reading.

The text, as printed herein may, therefore, be taken as fairly accurate. From the hymn (vv 22-35) it appears that whilst the worship of Vishnu and Shiva was popular, and their votaries were applauded, the Kulachara was blamed. A translation of this hymn in English was printed in the volume entitled "Hymns to the Goddess" by A.&E. Avalon. As was there pointed out, the text of the Tantrasara used for this translation was in parts corrupt and unintelligble and in others of doubtful meaning. A further translation with commentary has therefore been here made by A. Avalon of the text as it has now been revised; and the opportunity has been availed of to correct some errors.

Mahishamardini Stotra

1

O Chandi!

By Whom the act of the wicked and formidable Asura was shattered.

Do Thou wander in my heart.
Destroy my selfishness and the calamities which deeply pierce me,
Arising from the mass of malice and fears (which assail me),
So that, free from danger,
And protected by the lotus cluster of Thy feet,
My swan-like mind may swim and rejoice in the Ocean of Bliss.

2

What fear of enemies has he who worships Thee?
The Devas who worship Thy feet
Having abandoned the form of Nrisingha
Whose towering mane rivals in splendour and height towering Mount Sumeru,
And whose fingers are outstretched to tear (the breast of) Hiranyakashipu
Now worship (the lion) the enemy of the elephant
Server of Thy feet which destroy the bonds of the Pashu

3

O Chandi! when the syllables, the letters of which speak of Thee,
Reach the ear, then Brahma and other Devas
Sing the truth, touching Purusha and Prakriti.
O Devi! be today gracious to me,
Devoted as I am to the kissing of Thy sacred lotus feet,
The one and only glittering abode of the essence of the nectar of all Devatas.

4

If, because of my following Your way of Kula,
I suffer reproach, better is it that I shall thus be without fame.
Let me not have that which comes of the worship of Keshava and Kaushika;

Rather, O Mother! let my heart rest in meditation on Thy lotus feet,

Worshipped by Brahma, Hari and the Enemy of Smara
By the Eater of oblations and the Enemy of the Daitya.

5

O Mother! If I be engaged in the constant contemplation of Thy lotus feet,

Then what is there which Siddhas have which I have not

May Thy lotus Feet be ever present to my blissful mind

They feet from which exceeding mercy flows!

O propitious Mother! do Thou forgive me.

6

Verily and without doubt, even the Lord of Bhutas would have perished

Maddened as He was with the joy of the embrace of (Thee Who art) His own self

Had He not been freshened by the lotus fragrance of Thy feet,

Bathed in the honey which flows within

From the union of Shiva and Shakti.

7

O Mother! let the stream of heavy showers of holy devotion towards Thee

Be ever shed upon me,

Struggling and drowning, alas! as I am in the endless Ocean of Illusion

Without taste of the water of the Bliss of Brahman

Which devotion dispels the weight of anguish from numbers of Devas.

8

May the glory of Thy feet dark as a rainladen cloud,

Be every in my heart

Dispelling by its lustre as of ten million suns

The darkness which overspreads my mind.
From its glittering womb were born the three Devatas,
Who create, maintain, and destroy the world,
Whose substance is pure consciousness and bliss.

9

May Devi Durga Who gives victory and happiness
Dispeller of fear, Victrix of fortresses and ill-fortune
Who had power to destroy the proud enemies of the Devatas,
And Who strikes terror into the hearts of thousands (of Her foes)
Ever conquer!
She it was Who, having severed the head of the Asura Mahisha,
Crushed and killed under Her feet him who assumed the form of a buffalo
Now bellowing, now charging, and again retreating
And from whose mouth the Asura issued.

10

In the red ocean vast and surging
Danced the great shields, weapons and streamers of the enemy
(An ocean) clouded by the flight of discus
And the arrows of the heaving multitude of soldiers
There lay the heads of the proud and wicked Asuras
Broken and cut to pieces, tossed about by the storm of battle,
(The sight of which) sharpened the thirst and hunger of the birds of carrion.

11

I meditate upon Devi Mahishamardini,
Rushing in frenzy now here, now there on that wondrous field of battle (for the slaughter of the enemies)
Attended by eight companion Matris

And on the Mantra and Badhu Vija in the lotus of eight petals.

Within the two horns of the fierce and terrible restless and challenging head

Bent low and slanting of the maddened buffalo.

12

Let the Sadhaka meditate on the dark Shiva (Mahishamardini),

Holding in Her hands discus, lance, axe, shield, arrow, bow and trident,

Making the gesture which dispels fear;

Her mass of hair is like a bank of cloud entwined up on Her head,

Her face most formidable awes (Her foes)

Making even the defiant falter

Her laugh is loud and terrible.

13

O Devi! such as in this manner

Meditate upon this Thy faultless form

Or upon Thee as Durga or other form of Thine

Worshipped by Indra and other Devas,

To them it is given to attack the cities of their foes,

And conquering their enemies, to gain a kingdom;

They too, acquire the nectar of the knowledge of poesy,

And power to arrest, banish, and slay.

14

Whosoever reads or hears this Hymn

Made by me in rapt meditation upon Thy lotus feet,

Wherein is said Thy Kula worship and Mantra in hidden form

In the palms of the hands of all such

Are forthwith wealth, fulfilment of desire and liberation.

O Mother! salutation to Thee!

May Thou conquer!

In the concluding portion of this Chapter the Devi says that Her chief forms are represented by Mahishamardini, Kali and Tripurabhairavi, the last being considered the primary manifestation (VII, 37). This work inculcates the worship of Yoginis as a part of Kula-worship, on Kula-days (Kulavara) and Kula-tithis, especially on the 14th day of the moon.

[The Kulavaras have been described in the Yamalas from which they have been quoted in the Tantrasara as follows. "Tuesday and Friday are Kula days while Wednesday is noth Kula and Akula, the rest being all Akula". Again all Tithis with an uneven number are Akula; with an even number Kula with the exception of the second, sixth and tenth which are both Kula and Akula.]

The evening rite, consisting in the offering of food to jackals, is described at length. The Devi towards the end says "Thou art the Guru of all the Tantras and neither I nor Hari. Therefore Thou art the Revealer of the Tantras. I entered into Thy body (as Shakti) and thereby Thou didst become the Lord (Prabhu). There is none but Myself who is the Mother to create (Karyyavibhavani) and therefore it is when creation takes place that sonship is in Thee. Thou alone art the Father who wills what I do (Karyyavibhavaka) and none else.

Mang vina janani kapi naiva karyyavibhavini
Atah karyye samutpanne putratvang tvayi vartate
Tvang vina janakah ko'pi naiva karyya-vibhavakah
Atastvameva janako nastyanyo'pi kathanchana

"At times Thou art the father; at others the son; at times Thou are the Guru; at other the disciple. By the union of Shiva and Shakti creation comes (Shivashakti-samayogat jayate srishtikalpana). As all in this universe is both Shiva and Shakti (Shivashaktimaya) therefore, Oh Maheshvara! Thou art in every place and I am in every place. Thou art in all and I am in all."

Notes

1. A title of the Shakti of Shiva as the powerful victrix of demons. She is Mahishamardini, as the slayer of Mahisha. The Daitya Shumbha attacked Her in the form of a buffalo (Mahisha).

2. A form of the Devi assumed for the destruction of the Daitya Chanda, and who assisted in the destruction of the demon Raktavija.
3. Mahisha
4. Manohangsa; the Hangsa is variously described as a swan, gander, and flamingo.
5. The Man-lion incarnation (Avatara) of Vishnu, in which He destroyed the Daitya Hiranyakashipu, father of His devotee Prahlada.
6. The Avatara is generally represented with the King of the Daityas across his knees, tearing asunder with His hands and claws the latter's belly.
7. Which accompanies the Devi as Durga. After the destruction of Hiranyakashipu, Vishu's wrath was not appeased. The world trembled, fearing what He might do. The Devas asked the help of Shiva, who assumed the Sharabha form-that of a lion with wings and eight feet-who caught up Vishnu into the air and held him there until he had become powerless. The lion then went to the feet of Durga, whom he accompanies.
8. Kari the elephant form subsequently assumed by the Asura Mahisha. Karivairi = enemy of elephant = lion.
9. Shiva and Shakti; the "male" and "female" from whose union springs the universe.
10. That is Kulachara, one of the divisions of Tantrik worshippers, who, the verse says, are misunderstood, and therefore subject of reproach; and which is contrasted in the next line but one with the more popular and conventional worship of Keshava and Kaushika.
11. Vishnu
12. Indra
13. Vishnu
14. Smara, the God of Love; Shiva, who slew him, is his "enemy".
15. That is Fire.
16. Daityari; usually an epithet of Lord Krishna, but as

Hari has already been mentioned, possibly the reference may be to Indra. According to Medini, Daityari—Devata.

17. Siddhaspada.
18. This "mind" has a qualifying adjective, viz:—"Akshatasampadi", "of uninterrupted happiness." It is so because the Devi is dancing there.
19. Shiva is Bhuteshvara or Bhutanatha. Bhuta, which in a general sense means "beings," specifically refers to the Spirits by whom Shiva is surrounded, and of whom He is Master.
20. It is by the Devi's aid that Shiva is Parameshvara, for without Shakti he is nothing, and without Her life-giving energy and support cannot exist. As the Kubjika Tantra says: "Without their Shaktis the husbands are but Preta (inert corpses." So also the Jnanarnava: "O beloved, pure Sadashiva, without Shakti, is without motion like a corpse, for without Shakti He can do nothing."
21. Svatmanam parirabhya. Literally, having embraced himself. The Devi is, however, in a dualistic sense, His sacred half, and in reality one with Him and his own self.
22. Daivadvichyu-tachandra-chandanarasa-pralbhya-garbha-sravat. The meaning is not clear but Chandra (moon) seems to stand for the Vija of Shiva (which it also means): and Chandanarasa (liquid sandal flow) issues from Devi. Therefore the union of Shiva and Shakti in the Sahasrara appears to be indicated.
23. Mohajaladhi-vyahara-viddha, lit: "pierced by the mockery of the ocean of illusion."
24. See Markandeya Chandi vv. 38, 39.
25. Chamara.
26. Reading Smare for Vare in text.
27. The Devis so called *vide ante* the Karnejapa-Stotra.
28. String the Mantra may be said with this or the Tara (Ong) Maya (Hring) Kaama (Kling) or Vagbhava Vija (Aing).

29. On the petals are the eight syllables Mahishamardini svaha; "Salutation to the Devi slayer of Mahisha".
30. The buffalo when charging puts its head askew. The Mantra is thought of as placed between the two horns.
31. The Abhayamudra.
32. Stambhanam, Uchchatanam and Maranam; three of the Tantrik Shatkarma.
33. The Mantra Mahishamardini svaha may be spelt out from the first six verses from the following words which respectively commence them. Machchitte (Ma) Hitva (Hi) Chanditadvishayantara (Sha) Manninda (Ma) Nirddishto'smi (Rdini) Svatmanam and not atmanam as given in the text (Sva) and Haha (Ha).
34. *Ravichandrau guru saurish chattvarashchakula matah*

    Bhaumashukrau kulakhyau tu budhavarah kulakulah.
35. D*vitiya dashami shashthi kulakulamudahritam.*

    Vishamashchakulah sarvah sheshashcha tithayah kulah.

    Similarly all Nakshatras with an even number are Kula.

    Varunardrabhijinmulang kulakulamudahritam.

    Kulani samadhishnyani sheshabhany akulani cha.
36. That is She has the disposition to act or to carry out what He wills. She alone has Kartritta for the Father as efficient cause does not act, but the Mother in whose womb the seed of the world is sown alone does so.

Malinivijayottara Tantra

General: The supreme control over and the gradual unification' with the 36 principles from Shiva to the gross earth form, inter alia, the principal argument of the Malinitantram. The first recipient of this mystic lore is Paramesa who learns it from Aghora. The Tantram is in the form of an interlocution between the Devi and Paramesa, the sacred teachings whereof have been communicated to the Rshis by Kumara the destroyer of Taraka.

It seems from the opening verses that there was originally a Tantram called the Siddhayogisvaritantram which consisted of nine crore verses and which interpreted the Shaivaistic teachings in the light of Bheda, Bhedabheda and Abheda. Its abnormal prolixity necessitated its being summarised in three crore verses, the chief topic dealt with therein being the path of yoga.

The abstract thus formed went by the name of Malinivijaya. In course of time, another improved and still more summarised edition· of 12,000 verses of the same was adopted. The tantram in the present form is supposed to be the briefest resume of the above as far as it is most convenient and helpful to the dull brain.

In my opinion supported by the text itself, the Malinitantram is the latter part of the Siddhayogisvaritantram of which the former part was probably represented by the Siddhamatam (cf. sloka 13, M. V. T., 1st Adhikara).

Authorship of the Tantram: The Tantram is affiliated to the Agama Shastra and is, as such, of divine authorship. The origin of the sacred lore or Agama is very beautifully painted in the rich vocabulary of Tantraloka. It describes in a masterly style the way how it was revealed to the sages:

"The Logos, or the All-Transcending Word (Paravak), germinally contains all the Shastras or Agamas in their super-sensuous form. This Logos materialises gradually into the physical form as syllables or vocables and, as such, forms the units of speech. In its preliminary materialisation as Pasyanti, the two aspects of consciousness are totally merged together and the words and their meanings are not at all separated. The objects of perception in this stage appear as of a piece with the subject. In the intervening stage of the Para, words and their meanings are differentiated from each other in the mind before being voiced forth. In the last stage it becomes Vaikhari because it expresses itself through the words uttered by different vocal organs.

"Thus the Para, forms ultimately the source of all the Shastras flowing out as spoken words from the five faces of the Supreme Being, which represent His fivefold Shakti—Cit,

Ananda, Iccha, Jnana and Kriya. The five systems of revelation, which principally originate from Paramashiva through His five powers, constitute the 92 schools of Tantras divisible into three classes and designated respectively Shiva, Rudra and Bhairava. First of these which maintains the doctrine of duality consists often systems only.

The second class contains eighteen systems and propounds the doctrine of duality-unity or Bhedabheda. The last, on the other hand, is unique in declaring the eternal truth in the light of Idealistic Monism and preaches its doctrine in 64 systems."

The same line of thought is also perceptible in the primeval utterances of our ancient vedic seers. The reader is advised to consult the verses 29, 45 hymn 164, mandala I, Rig Veda.

1. ayam sa shimkte yena gaurabhIvrita
   mimati mayum dhvamsanavadhishrita.
   sa chittibhirni hi chakara marttya.
   vidyudbhavanti prati vavrimaudata..
2. chatvari vakh parimita padani
   tani vidurbrahmana ye manishinah.
   guha trini nihita nemgayamti
   turiyam vacho manushya vadanti..

Tantra Defined: Tantra being a generic term for a class of religious and magical works is designated as the mantrasastra though it generally treats of (i) incantations, (ii} their philosophy, (iii) the principles, (iv) the worlds, (v). ceremony, (vi) initiation, (vii) worship, (viii) mental and bodily discipline, (ix) observances of a Tantric. Like the Veda it is, as the Tantrics hold, the breath of the Supreme Being or in other words the immutable and eternal knowledge.

The performance of the present day Vedic ceremonies by the Kashmiris, which run side by side with those of the Tantras and which are sometimes intermixed with the latter, leads to believe that the Tantras are either an imitation or a development of, or an improvement upon the Vedic ceremonies. The Tantric supplement is useful, in the opinion of the Tantrist, to quicken and ensure the fulfilment of desires in the Kali age.

Tantric Worship: It does not essentially differ from that of the Vedas. It strongly maintains the Omniscience and the omnipotence of God. The complete immersion of the duality in unity is the main lesson of the Tantras. The adorer becomes the adored himself. His body is the temple and his soul the idol in it. He is the high priest not of a god above; around, or below but of his own ever-wakeful self.

Tantric view of Moksha: Moksha, in the tantric sense of the word, is the unfoldment of powers brought about by the self-realization. It is not the giving up of the mortal coil and thus acquiring immunity from death. To a real Tantric, birth and death are phenomena of his own creation. He finds gratification as much in the one as in the other. He is the Bhairava whose name even strikes terror into the destructive agencies and at whose sweet will the world lives, moves and has its being. His sole article of faith is that death has significance only for those who are subject to mortality and not for those who have risen superior to its idea by their living belief in the deathlessness of the soul.

Realistic Optimism of the Tantras: A Tantric, like the one believing in the Vedas, by gradual marches on the spiritual path, comes to feel his close kinship with nature. To him both the animate and the inanimate are glowing with divinity. The divinely glorious presence puts an end to all his animal tendencies and inspires him to fulfil his noble mission. His self-centredness cheers him up and infuses him with the fresh energy that facilitates and quickens the dawn of his native glory without any detriment to the enjoyment of the sweet pleasures awaiting him in the world. To him the world is not an illusion, not a Maya of the Vedantists, and therefore he tries to make himself and his surroundings happy to the best of his ability.

Potency of the Mantras: The main theme with which the Tantras are concerned is the power of Mantras. Mantras, as the Tantrists maintain, are certain scientifically arranged formulas which, if practised according to the Tantric precepts, bring about certain results conducive to the fulfilment of utterer's wishes. Each and every Matrika (a, A, etc.) is a living energy in itself and should in no way be mistaken for a mere Varna or letter. A living force is created by placing the letters in a certain systematic order by the Sadhaka.

The mantras are nothing but the harmoniously living forces strong enough, no doubt, to accomplish even the most difficult undertakings of the votaries of the Tantras. This is not all. The arrangement of letters in a scientific procedure goes a great way to help the worshipper, provided no omissions of any sort of the part or parts in the Tantric ceremony are made, to fathom out the hidden secrets of nature, nay work wonders, and gradually free him from the meshes of the world and raise him high to the lofty plain of supreme consciousness styled Paramapada.

All-embracing Nature of the Tantras: The great merit of the Tantra lies in its all comprehensive and all-embracing nature. The portals of Tantra are invitingly open to all irrespective of caste and creed. The repulsive and revolting creed of untouchability is unknown here. The sweet cup of nectar, which it holds in its outstretched palms, can be enjoyed by both the high and the low. No hard and fast rules, nor any strict observances like those of the Vaidikas, are to be followed by the devotees of Tantra. Short-lived and enervated as the men of iron age are, they, by following the short and smooth methods of Tantric self-culture, accomplish within a short span of time what others achieve in ages after ages. This is why the Tantras are popular and have a large following.

Difficulties in the adoption of Tantric methods—But no rose is without a thorn. Good and evil run side by side. There is nothing which has unmixed good or unmixed evil in it. The rosy path of the Tantric is beset with thorns here and there. It is therefore the foremost duty of the Sadhaka to avoid the thorns and to adhere to the rosy path, lest he fall into the trap of the low spirits of nature, who, trying to cause mischief, put obstacles in his way and thereby retard his progress in the attainment of his desired object. It is only the potency of the mantras that brings about the subjugation of these supersensuous beings.

But, as ill luck would have it, a novice or a beginner in the Tantric ceremonies, oftener than not, associates with the low spirits and makes them minister to his unlawful ends. He discards the achievement of the high ideal which he ought to have followed strictly and unerringly.

This sort of practice is resorted to by the black magicians whose fate is rendered miserable both here and hereafter. Practices of this type are, never and in no region, commendable on the part of the real Tantrics. The realisation of the Advaita Tattva is the chief aim, or the summum bonum, to be attained by the high-class Tantrics. It is to this end that they consecrate their lives and gain the highest status in spirituality.

Malini: In his commentary on the 17th verse of the Tantraloka, volume I, p. 35, No. XXIX of our Series, Jayaratha says that the Malinivijayottara is so called because, due to the glory of Malini, it occupies a pre-eminent position among and is superior to all the other Schools of Shaivaistic Tantrism. Malini is also known by Mala and so the Malini is the generic term for the series of letters or the alphabet.

The natural and phonetically scientific order of the Sanskrit alphabet goes by the name of Purvamalini which is popularly known as Matrika or Siddhi. When, on the other hand, no attention is paid to the alphabetic origin and to its natural order and when therefore the vowels and consonants are taken promiscuously, the alphabet is called Uttaramalini, chiefly denoting that order in which 'na' occupies the first position and 'pha' the last.

Malinivijayottara has reserved for itself the various modes of application of the Uttaramalini for the mystic practices just as the Svacchanda does of the Purvamalini. I believe that in the time of the Sri Svacchanda either the order of the alphabet from 'na' to 'pha' was unknown or less appreciated and therefore less in vogue, or that the followers of the Svacchanda did not believe in the Malinimata. Of course Kshemaraja looks at the Svacchanda in the Trika perspective but there is no reference whatsoever to the Malini, *i.e.*, the Uttaramalini, in the text of Svacchanda.

The Sanskrit alphabet as given in the Uttaramalini arrangement runs as below:

na ri ri li li tha cha dha i na u u ba ka
kha ga gha na i a va bha ya da dha tha jha ja ra ta
pa cha la a sa ah ha sha xa ma sha na ta e ai
O au da pha.

Malini is of the greatest utility in infusing the divine life into the body of practisers and all the Tantrists are enjoined to have recourse to it for the attainment of desired objects in all the cases in which particular details and instructions could not be punctiliously observed regarding nyasa. Different parts of the human body represent the different letters of the Malini. While giving the three kinds of mantra called para, apara and parapara, instances are furnished as to how a particular incantation can be made out from particular technical terms such as head, eyes, etc.

Commentary: The text of the Malinivijayatantra is difficult to be understood owing to the numerous technicalities found in it. But the attempts of Abhinavagupta have left no point unexplained and for this the Tantric public must feel specially thankful to him. The first verse of the Tantram he took up in his Malinivijaya varttika and explained it so thoroughly that connotations of several terms are made clearly intelligible. The points, that could not find room in the Varttika for want of space and owing to its limited scope as Varttika, he discussed in the Tantraloka.

The Tantraloka, as the author himself declares, is a detailed exposition of the Tantram. The reader of the Tantra could get an opportunity of understanding the text more clearly if this Department were to get the MS. of Abhinava's Purvapanchika on it. Abhinava has a very high opinion of the Tantra and says in the Tantraloka that the Malinivijaya is the essence of the Trika System which again represents the pith and marrow of the different teachings of the Lord enunciated in the different schools of Tantras.

Abstract: The great sages Narada and others, feeling athirst under the holy inspiration of Shivashakti for the right realisation of the Supreme Principle, repair to Kumara and, after paying due homage, request him to initiate them in the mystic processes of yoga. Kumara, thereupon takes commiseration on them and communicates the secret knowledge of the Malinivijayatantra as originally addressed by Shiva to Uma.

The whole range of the knowable is divided into two classes, the acceptable and the avoidable. The acceptables are: Shiva, Shakti, Vidyesa, Mantra, Mantresvara and the Jivas. The impurity, actions, Maya and the whole world, as her creation, form the category of the avoidables. The key to the success in both the spheres of matter and soul is supplied by the right discernment of the acceptable and the avoidable.

The supreme controller of the universe in the words of the Malinivijaya is all-doing, all-knowing, all-sustaining and infinite. On the emergence of desire at the creational stage, He evolves out of His own Self the octad of Vijnanakevalas. He assigns to them the functions of sustenance, destruction, preservation and benefaction. For their sphere of action He manifests seven crores of mantras with their respective ranges. All these mantras are the living psychical forces and they gratify the wishes of a Sadhaka by the well-merited bestowal of fruits.

The self comes into manifestation in the fourfold capacity as Shiva, Mantramahesa, Mantresa and Mantra. Shiva is the lord Himself. Vijnanakala is a degree less than Mantra owing to the defilement of mala. Pralayakala is wrapped up in the defilements of mala and karma. Mala stands for the imperfect knowledge and the tree of the world shoots up therefrom. Karma is the action in general. As virtuous, it leads to pleasure and, as otherwise, it yields pain. Sakala stands for the anus or jivas in general. Their field is offered by the totality of categories from kala to earth.

Rudras are one hundred and eighteen in number and are headed by Angushthamatra.

According to their merits they are appointed as Mantresvaras. They take over charge of their departments and reward Brahma and others according to their deserts. The sages receive the sacred science of the knowable either as acceptable or as avoidable from Brahma, etc. and Manus, from the sages.

Of the seven crores of mantras one half of the portion has, at the option of Shiva, attained the deathless stage after favouring hosts of jivas.

Each principle from earth up to matter is susceptible of being viewed from fifteen different standpoints owing to the seven perceivers regarded either as Shakti or as Saktiman and to the principle itself; those from Purusha to Kala, from thirteen owing to the inapplicability of perceptivity to Sakala therein. The principle of maya is viewed from eleven standpoints because of the further reduction of Pralayakala as.perceiver. Similarly, other principles know each a further reduction of two standpoints up to the last principle of Shiva which has no diversity.

The spiritual teacher is defined as one who knows all the aforesaid principles in their true perspective. He occupies as lofty a position as Shiva and is, as such, the revealer of the potency of mantras. At his mere touch the people get purged of all sins.

The inspiration of Rudra Shakti is ever existent in him. The five indicative marks of the Rudrasakti are: (1) firm devotion to Rudra, (2) the success of the mantra capable of yielding the desired object instantaneously, (3) the control over all creatures, (4) the bringing to completion of all undertakings, and (5) the poetic faculty. The inspiration of Rudrasakti is primarily divisible into three heads, Anava, Sakta and Sambhava, each of which is fiftyfold with reference to bhuta, tattva, atman, mantresa and Shakti and to their Sub-divisions.

Anava is that inspiration of Rudrasakti which is attainable by utterance, conscious mental discipline (karana), contemplation, incantation and the bodily centres.

Sakta is the name of that stage which is obtainable by the contemplation through the mind on the object of meditation accompanied by utterance.

Sambhava indicates the state which is produced by the deep and penetrative spiritual insight involving absolutely no mental concentration on any particular object.

All the above forms of inspiration are susceptible of being interpreted in the light of the fivefold state, waking, dreaming, dreamless, the fourth and beyond the fourth. The following table is given to help the reader in tracing the pentad of states in the chain of standpoints undergoing a gradual reduction by twos at each stage of perceivers:

From Earth to Matter

1. Svarupa-sakala stage
   - (a) Svarupa-sakala-sakti = waking state (jagrat)
   - (b) Pralayakala = dreaming state (svapna)
   - (c) Vijnanakala = dreamless state (sushupti)
   - (d) Mantra, mantresa, mantramahesa = 4th state
   - (e) Shivashakti = beyond the 4th state

From Purusha to Kala

2. Sakala stage
   - (a) Sakala = waking state
   - (b), (c), (d), (e) as above.

Principle of Maya

3. Pralayakala stage
   - (a) Pralayakala = waking state
   - (b) Vijnanakalasakti = dreaming state
   - (c), (d), (e) as above

Mayordhva

4. Vijnanakala stage
   - (a) Vijnanakala = waking state
   - (b) Mantra = dreaming state
   - (c) Mantresa = dreamless state
   - (d) Mantramahesa = the 4th state
   - (e) as above

Suddhavidya

5. Mantra stage
   - (a) Mantra = waking

(b) Mantresa = dreaming
(c) Mantramahesa = dreamless
(d) Shakti = the 4th
(e) Shiva = beyond the 4th

Isvara

6. Mantresa stage
   (a) Mantresa = waking
   (b) Mantramahesasakti = dreaming
   (c) Mantramahesa = dreamless
   (d), (e) as above

Sadashiva

7. Mantramahesa stage
   (a) Mantramahesa = waking
   (b) Kriyashakti = dreaming
   (c) Jnanasakti = dreamless
   (d) Icchasakti = the 4th
   (e) Shiva = beyond the 4th

Undifferentiated Stage

Shiva Stage

(a) Kriya = (a) Waking
(b) Jnana = (b) Dreaming
(c) Iccha = (c) Dreamless
(d) Ananda = (d) The 4th
(e) Cit = (e) Beyond the 4th

The synonyms of the states are:

(a) Jagrat = Pindastha = Sarvatobhadra.
(b) Svapna = Padastha = Vyapti.
(c) Sushupti = Rupastha = Mahavyapti.
(d) Turya = Pracaya = Rupatita.
(e) Turyatita = Mahapracaya.

The triple manifestation of the transcendental self as an average self, as energy and as Shiva, constitutes the triad of fundamentals forming the central theme of the Trika Philosophy. It covers the entire field of Tattvas of the Shaivaism, the Atman overlapping as many as those beginning with earth and ending with the Vijnanakala; Vidya or Shakti, those from Mantra to Mantresvara and the rest constituting the state of Shiva.

The fourfold classification of the Tattwas from the evolutionistic standpoint is termed the earthly, the material, the mayic and the Saktic. Among these the first is pervaded by Dharika Kala. It has only one Tattwa, one letter, one world, one mantra represented by 'ksha' and sixteen worlds. Apyayani Kala pervades the second. Twenty-three principles from water upwards and as many letters (ta and others) are assigned to it. Five words, five mantras and fifty-six worlds are included in it. The third is permeated by the Bodhini Kali. Seven principles, twenty-eight worlds, seven letters, two words and two mantras are found in it. The last is covered by Utpuyini and there are three principles, three letters, eighteen worlds, one word and one mantra in this class. The final principle is called Shiva, as such, it is under the Avakasada kala and has sixteen vowels, one mantra and one word. The above four classes are respectively presided over by Brahma, Vishnu, Rudra and Isvara.

In giving the genesis of the Sanskrit alphabet from 'a' to 'ksha' there occurs a description as to how the inseparable Shakti of the Lord, at the cosmic stage, evolves as cognition and activity when He feels the holy impulse of bringing into manifestation the world as we see it. The faculty that gives the determinate knowledge, 'this is thus and not otherwise', is known as jnanashakti. The kriyashakti denotes that faculty wherethrough the determination, 'let such and such thing come to light', is translated into action. As regards the universe of sound, Shakti appears as Purvamalini representing the alphabet from 'a' to 'ksha'. Aghora is said to be awakened into action by Paramesvara and at His desire he vitalises maya and from her emanate various letters and these letters are the source of knowledge in general.

Although the Shakti of Shiva possesses infinite forms, she is chiefly known through her three aspects—(a) apara or ghoratari, (b) parapara or ghora, and (c) para or aghora. The apara keeps always surrounding the Rudra souls and pushes the Jivas on the downward march by attaching them to sense objects. Parapara, like the former, hinders the progress of the jivas towards their goal of life and brings about their attachment to the mixed fruit of pain and pleasure. Lastly, para occupies herself in conducting the creatures to the highest end of life, the attainment of Shiva stage. The Uttaramalini in which also Shakti appears is referred to above.

The three respective mantras of the above triad are given in the phraseology of the different parts of the human organism.

The allotment of the six systems of the path (adhvan) is substantially the same in the Uttaramalini as in the Purvamalini. The former in this respect differs from the latter inasmuch as the order of the alphabet is different to and other than that given for the former. 'Pha' is assigned to the principle of earth. From 'da' to 'jha' the twenty-three 'letters are respectively allocated to the corresponding 23 principles beginning with water and ending with matter. The principles from purusha to maya claim one each, the letters from 'cha' to 'a'; 'i' to 'gha' cover respectively, the principles Suddhavidya, Isvara and Sadashiva. The principle of Shiva embraces the sixteen letters from 'ga' upto 'na'.

For understanding the application of the three mantras of para, apara and parapara, as regards the different principles, the reader is advised to consult the passage given on page 24 Malinivijaya.

The knowledge of the acceptable as also of the avoidable is indispensable for the attainment of yoga which, as such, is the unification, of the individual soul with the universal. Jnana as well as yoga lead alike to the goal of life as referred to above. The former is threefold as being based on what is heard, on contemplation and on living realisation of the reality. The first deals with it as dwelt upon here and there in the sacred texts. The second involves a deeper study of those in their entirety with an eye to sifting the fundamentals. The third marks the final stage of the seeker after truth and is possible only on the fixity of the previous stage.

Yogi also has to go through the four stages while on the path of Yoga. According to the degree of divine inspiration, he is called (1) Samprapta, (2) Ghatamana, (3) Siddha and (4) Siddhatama.

Samprapta is one who has received initiation through a line of teachers and is living upto it. Ghatamana refers to him who again and again checks and restrains his mind from the outside influence and. concentrates it on the truth.

By the continuity of efforts owing to the ever-increasing interest, he gets success in disciplining his mind into the concentration on that truth exclusively, and is called Siddhayoga.

When getting totally identified with the highest principle so as to lose his individuality to the utter disregard of the circumstances he is placed in, he becomes siddha. In this stage alone he is really justified to uplift the depressed humanity.

The shaiva teacher is advised to purge his disciple of all the impurities peculiar to the different worlds numbering 118 which respectively cover the range of principles from Prithivi to Sadashiva before he can lead him to find the reality in Shiva.

The purification of the soul is to be necessarily undergone not only in the above way but also in the body which microcosmically represents all the principles.

The Malinivijaya has a set of mudras on the exercise of which it lays a great stress and enumerates them as follows: (a) Trisula. (b) Padma. (c) Shakti. (d) Cakra. (e) Vajra. (f) Danda. (h) Mahapreta. (i) Mahamudra. (j) Khagesvari. (k) Mahodaya. (l) Karala. (m) Khatvanga. (n) Kapala. (o) Hala. (p) Pasa. (q) Ankusa. (r) Ghanta. (s) Trisikhamudgara. (t) Avaha. (u) Sthapana (v) Rodha. (w) Dravyada. (x) Nati. (y) Amrta, and (z) Yogamudri.

Their mantras are to be formed by prefixing the syllables 'om' and 'hrim' and affixing the word 'namah' at the end.

A follower of the Malini School, if and when he desires to perform a sacrifice, must have purificatory baths technically known as Bhasmasnana, Jalasnana, Agnisnana, Vayusnana and Divyasnana. Being thus purified he should enter the sacrificial altar and offer worship to the deities of the entrance. He should consider himself as akin to Shiva in the form of light and should electrify himself with the vital energy of consciousness consisting in 'I am He.' He should identify his body with the matrika assigning each of the parts of his body to the corresponding letters of the alphabet according to the arrangement of the matirkanyasa. In the saktanyasa which he has to observe after this, he has to, identify his body with the three vidyas alluded to above. Then he has to perform the mental sacrifice for the details of which the student is advised to consult the text p. 48 Malinivijayatantram.

As regards the physical worship the worshipper is to worship Ganesa as three-eyed, as elephant-headed and as dwarfish in stature. The puja to Ganesa being over, he has to contemplate Shiva as seated on the sixfold seat of Ananta, Dharma, Jnana, Vairagya, Aisvarya, and Karnika and as bearing a sword, a shield, fierce to look at, of great jaws, terrible and with his eyebrows knit together and surrounded by the eight deities (matrikas).

Then he should begin the sacrifice. Two vessels, one big and one small formed of gold and full of perfumed water, should be placed with two canopies of white cloth over them. Indra and other deities of the quarters should be requested to watch the proceedings so as to ward off the malignant influences while the smaller vessel, is passed round and its water allowed to drop. Then he should prepare a kunda for the fire. The fire should necessarily be got either in a copper vessel or in an earthen pot and should receive the several ceremonies of birth, etc. He should perform homa by muttering the mulamantra one hundred times and the other attendant mantras ten times The homa being over, he may rest and go to sleep.

If he sees a good dream he may express it to his disciples and if otherwise he should perform the homa.

The vows that a Tantric disciple has to observe are: (A) that he should always offer worship (a) to his deity, (b) to the fire, (c)to the spiritual teacher and (d) to the goddesses; (B) that he should not eat without offering his food to them;

(C) that he should refrain from using the property dedicated to the deity, the guru and Candi; (D) that he should not even for a moment engage himself in idle and unavailing pursuits and lastly (E) that he should either be fixed in the practice of yoga or in that of the mantra.

When by the proper and satisfactory discharge of the duties prescribed for him, the disciple gives an unmistakable proof of his ardent admiration for the Shaivism and when consequently the Shaiva teacher also feels justified in initiating him in its mysteries by his own conviction in the depth and genuineness of the disciple's longing for the same, the latter should have the mystic diagrams drawn on the carefully selected piece of ground in conformity with the rules and regulations detailed in the Malinivijayatantram. In their centre he should invoke Bhairava and perform the puja to Him. Thereupon, he should approach Him with the following:—'At Thy dictate, O Lord, I have been installed in the position of a preceptor. The disciples are fully under the sanctifying influence of Shivashakti. They, as such, deserve Thy favour. They have sought Thy help in securing it. Therefore, O Lord, I pray, be kind to me and let my body be enshrined by Thee so that I may be qualified to render them due assistance.'

The prayer being over, he should believe and feel that the six kinds of the Path are presided over by Bhairava and that his body has veritably been penetrated into by the divine spirit and the cloaks of mala that conceal his kinship with Shiva have fallen one by one laying it entirely bare to him. In this stage he should identify himself with his disciple and with the path in which the disciple is to be initiated by the holy faith, 'I alone am the supreme reality; this whole universe is in me; I am' the stay and support of this all.'

The unification of the Sishya, the teacher and the rest of the world into the one ultimate reality is the Nirvana. This is as regards the outer aspect of initiation.

As regards the inner side, the teacher has to bring in the Jiva of his disciple from the outside into his own self and then he has to awaken the serpent force in him and move it up from centre to centre till it reaches the Brahmic aperture in the head. The disciple, that is thus united with the supreme principle by his own spiritual teacher, witnesses the destruction of his bondage of maya and never, thenceforth, takes his birth and becomes a pasu. If the disciple after receiving initiation feels: a strong impulse to acquire the mystic powers, he may repair to his guru and obtain instructions from him regarding the performance of certain penances leading to that end.

In case the Sadhaka aspires after the attainment of the stage of Acharya, he must acquire a higher mode of initiation leading to enjoyment both here and hereafter. The formation of the diagrams and the kundas and such· other external requisites have no significance in it. He has only to consider and really believe that his body has been burnt by the bright fire of energy enkindled from the tip of the toe up to the cranium through the help of mahamudra. After that he has to consider himself as possessed of an ethereal form purged of the impurities closely accompanying the principle of materiality.

The inspiration of Shivashakti as revealed in him is determined by the particular movement of Dantakashtha. The reception of the divine inspiration is betokened by the five states of (1) happiness, (2) the awakening of the serpent force, (3) the bodily tremulation, (4) sleep and (5) intoxication. In this initiation sixty-four lamps are to be lighted and the worship to Shiva is offered in a conch full of herbs and perfumed water with which the Sadhaka is anointed towards the end of the ceremony.

A spiritual teacher, who has already obtained control over the different principles constituting the world, must very well ascertain the tendency of the disciple towards a particular principle, before he instructs him in the way to achieve success in securing the conquest thereof. He should never allow him to swerve from that with which he has made an effort already to unify himself. He is sure to get union with Shiva through that after enjoying particular privileges peculiar thereto.

But the place must invariably be in all cases very well chosen for entering upon the yogic practices. It must either be a cave or a cell with no din of the world and perfectly charming as regards scenery. It must be a safe retreat from all evil contaminations.

Such a place very well serves the purpose of the yogi who has habituated himself to a particular posture day and night. A man of the street or a man of the field cannot take to yoga and profit by that. A candidate for that must have curbed his passions to the entire subjugation of the mind. He must have had an established practice in the suppression of breath. All the sense activities must be under his control. He must have overcome sleep, anger and the mental restlessness and must be altogether impervious to pain.

A candidate with such qualifications is entitled to yogic practice and if he continue to follow the methods adopted for different dharanas, he is sure to win his conquest of the entire cosmos. The palm of the glory attendant upon the success in each dharana from earth to Shiva is very well described in the latter portion of the Malinivijaya. If the seeker after truth dives deeper in the mysteries of them, he will himself examine the truth of the statements made so often in connection with the dharanas. His patience will not be exhausted by the minute details recorded in relation to the dharanas because the success in one particular dharana will repay the trouble in an incalculable way.

The Netra Tantra

The *Netra Tantra* is attributed to the school of Kashmir Shaivism and is usually printed with a commentary by Kshemaraja.

The work, divided into 22 adhikaras of uneven length, centres around Mrityunjaya, here described as Amritesha, and his cluster of shaktis. Chapter one is below. In chapter two, there is discussion of the three shaktis Iccha, Jnana and Kriya. Chapter three is concerned with the puja (yaga) of Mrityunjaya, while chapter five discusses initiation (diksha). In chapter six, the rules of abhisheka are given, while chapter seven deals with Amritesha, or Shiva as god of amrita or nectar. This has gross, subtle and supreme meanings, the text says.

Chapter seven is interesting because it is close in nature to the *Siddhasiddhantapadhati*, enumerating the six chakras, the sixteen adharas, the three lakshyas and the five vyomas (aethers), as well as the granthis and the nadis within the body.

Chapter eight deals with yoga of the tantras, while chapter nine begins to discuss the different tantrik divisions known as vama, dakshina, siddhanta, Saura and Vaishnava, and how they relate to the Vedas. Amritesha, says Bhairava in the text, is pure, like crystal and extends everywhere, giving the fruit of all agamas (sacred tets). The chapter gives different meditation images of Shiva.

Chapter eleven, devoted to the Uttaramnaya, starts with a dhyana of Tumburu, who is of the colour of dazzling white snow, or the kunda flower. Shaktis mentioned in this chapter include Jambhani, Mohani, Subhaga and Durbhaga. Chapter twelve concerns the Kulamnaya, and outlines the mandala of Bhairava and different shaktis and worship conducted there. More meditation images are contained in chapter thirteen, which also contains a rare dhyana of Brahma. This teaching is open to all, be they female, male and of whatever caste and hue. Brahma is described as having four arms, handsome, red in colour, effulgent, seated on Hamsa (a swan, but here meaning the mantra). He holds a staff, a rosary of akshas, a jewelled water pot and the four vedas.

In chapter fourteen, the role of this mantra and Iccha, Jnana and Kriya Shaktis are discussed, and the supremacy of the mantra. Chapter fifteen describes how Amritesha's mantra is all protective, while the next chapter describes different siddhis obtained from the worship. That topic is continued in chapter seventeen, which also covers the kavacha.

Chapter eighteen is devoted to Amriteshvari, or the shakti of Amrita, as well as describing the purifications that mantras must receive to become successful. Chapter nineteen is long, with 226 shlokas (verses). It starts with the Devi asking Shiva to describe afflictions caused by bhutas, pretas, yakshas, pishachas, rakshas and the like, and how they can be prevented. Chapter twenty deals with the yoginis, such as Shakini and others of the bodily dhatus. Chapter twenty one discusses the nature of mantra, while the last chapter concludes with the great merit of Amritesha (Mrtyunjaya's) mantra.

Chapter One

Hail to the ordainer of destiny, the being who manifests three ways in the three worlds, the possessor of Shakti who creates, maintains and destroys in the cosmos, the being whose nature is amrita, Shiva, the supreme essence of Brahma, Vishnu and Isha.

Seated on Kailasha Peak is the god of gods, Maheshvara, Hara, the altar of dalliance, with his hosts and his spouse Parvati.

Having seen the happy god, and with the desire of benefiting living beings, suddenly Parvati left his side, and grasping his feet, questioned the contented Parameshvara in a very devoted way.

Shri Devi said: Lord god of gods, Lokanatha, lord of the cosmos, you have accomplished a great miracle, a cause of astonishment. You are god of all that exists, but my supreme master.

This secret, hard to distinguish and difficult to accomplish, is unknown to Kartikeya, to me, to the gods or to the ganas. It is certainly unknown to lords of yoga, to the Matrikas, to the rishis and to the yogis.

Lord of creation, speak now of this, if you are kindly disposed towards me! O Lord, I entreat you by your obligation to speak fully.

Thus having heard the words of Devi, the one with the smiling face spoke: Ask anything you wish, O one with beautiful hips. The secret is in your heart. I will certainly speak fully. You please me, O pure one!

Devi said: Bhagavan, lord god of gods, cause of various miracles, beautiful one of miraculous appearance, I wish to hear of that not already revealed. Handsome lord, I want to hear about the cause of the utmost bliss, to be related to me by you.

The all-seeing eye is made of water, Deva. How, then, may it become fiery and wrathful, flaming and burning up time? Saturn was reduced to ashes by the power of this eye. Deva, how is such wrath produced, that fire which desires to burn time? It consumes all creation, destroying Brahma and all that is permanent.

In a similar way, Parameshvara, Kaama was burned up by its play. What is this cruel, fiery eye, O Natha, which is always invisible yet is the cause of great miracles? How does fire come to be within this eye? Who does it see? How may an eye be made of fire? Why is it invisible? O cosmic lord, how comes it that this eye, the essence of immortality, augmenting the whole cosmos, has given birth to the cosmos?

Deva, these graceful nectar-like eyes are the cause of my bliss and behind the process of creation. How may this fire known as the fire of time come to create? Bhagavan, I want you to answer all this.

Shri Bhagavan said: I am moved by the great eagerness of your questions. Listen, dearest, I will speak of all relating to the fire and ultimate nectar which is within my eye and of its yoga.

Its real nature is without origin, pure, pervading all and omnipresent. It is within all living creatures and present in the hearts of all things, attained by yoga, difficult to accomplish, hard to attain for all beings.

It is like my own semen, self-knowledge, my supreme part. It is the essence of all semen, the strongest of the strong. Certainly, and without any doubt, it is the quintessence of all ojas, eternity itself.

From me came she known as supreme Iccha Shakti, one with Shakti, born from my own nature. Just as fire and heat and the sun and its rays are inseparable, so also Shakti herself, the cause of creation, is inseparable from the cosmos.

Within her is that which is both manifest and unmanifest. She is all-knowing, with all qualities, manifested as Iccha, Jnana and Kriya and so forth, and in her, knowledge, the six qualities and everything else are situated. All light dwells in her.

She is the essence of Mahakriya, the unified mother of action. She both creates and destroys and is the very self of Anima and the rest of the eight siddhis. Thus, these three Shaktis of mine are called Iccha, Jnana and Kriya, it is said. In me dwell the three playful abodes of Sun, Moon and Fire. In the play of my magnificent three eyes is the substance of these three. I create, sustain and destroy the universe.

I am the dwelling place of the three bodies and of creation, maintenance and dissolution. My effulgent and life-giving semen pervades all. With my forms of Iccha, Jnana and Kriya I am the ultimate eye nectar.

This semen is the supreme realm, the highest form of nectar, supreme bliss, the quintessence, complete knowledge, pure, the core of the three eyes. This is called the Mrityunjaya (conqueror of death) and gives success to all. He (Mrityunjaya Shiva) is the giver of success, the supreme divinity, liberating from all sorrows, the god destroying all ailments, removing all delusions, Shiva, the alleviator of poverty, eternal, conqueror of death, permeating all, infallible, without stain, peaceful, all-giving, all-liberating.

His brightness is equal to 1,000 million suns and 1,000 million fires, liberating from the sixteen kalas, effulgence itself, unassailable by gods or demons. With my fiery eye I burn everything in an instant and I may also create and maintain. There is nothing greater than this certain semen-like thing seen everywhere, the essence of vajra, taking one to the state of Rudra, like a renowned sword which is death to all enemies and stops all elementals, weapons and arms.

This one semen becomes multifold, diffusing itself limitelessly with many variations. The magnificence of this great Pashupata is that it is like Vishnu's discus or Brahma's staff and is the very essence of all weapons. Appearing in various forms, this weapon spreads in many ways. My semen creates the different gods themselves.

I, the lord of yoga, through my own Shakti, manifested the entire cosmos. She is the supreme protectress from fears and anxieties, allaying fear, destroying enemies and the supreme giver of liberation, most certainly. O Beautiful One, even great poetry could not describe the greatness of this!

This great thing, the giver of grace, the most excellent Boon-giver, causes manifestation, maintenance and the great intensity of Rudra. It should be regarded as immeasurable, knowledge itself, the great power of mantra, the protector of all the elements. Very hidden, you should always conceal it. Devi, it has now been revealed to you. What else do you wish to ask?

Niruttara Tantra

This is a relatively brief work of 15 chapters, belonging to the *Kali* class of tantras and written in readable and simple Sanskrit. The word Niruttara means "having no better".

Chapter one deals with the three bhavas or temperaments of a tantrik and describes how the different amnayas, a term which here refers to the five directions (north, south, west and east including upper) relate to the classifications into divya (divine), vira (heroic) and pashu (herdlike). Verse 16 gives some definitions: "A divya is one in whom devata predominates, while a vira is strong-minded (uddhata). The acts of a pashu relate to the Eastern Amnaya it is said." Uddhata, which is translated here as strong-minded, is actually a technical term of the tantriks—it means a person in whom the rajas or active guna predominates.

The best 'directions', according to this work, appear to be the Northern and the Upper, both of which have the characteristics of vira and divya bhavas. Vaidika worship is for the day while Kula acts are performed at night. A vira should not worship during the day while a pashu should not worship at night-time.

The cremation ground is declared to have two meanings. One is the place 'where corpses sleep' while the other is in the form of the yoni. But the worship must be dual. Without Kalapuja, which in this context apparently also means puja of the lingam of Shiva-Mahakala, Dakshina Kalika does not give fruit.

Chapter two begins a description of Dakshina Kali, her mantra, her preparatory acts (purashcharana) and the results it gives. Shri Shiva says: "One should know that the vagina (bhaga) is Bhagavati, she is Dakshina and the lady of the three gunas (Triguneshvari). This vagina-form is all, that which moves and that which does not move."

At the centre of the yoni, which also here means a downward pointing triangle, is the Ha-ardha kala which is the subtle form of the Devi. The yoni is Dakshina Kali and she is the essence of Brahma, Vishnu and Shiva. When semen is in the yoni of the devi, she becomes Mahakali, the form of light, and gives birth to the universe. Shiva and Shakti are of two kinds, with qualities and without qualities. Without qualities they are a mass of light, the supreme absolute (Parabrahma), eternal. In reverse intercourse (viparitarata), Kali is both with qualities and without qualities. When she takes the form of the new moon, she is without qualities and is known as Aniruddha Saraswati. When associated with Vishnu, she takes the form of Mahalakshmi, and is Maya herself. In her form as Dakshina Kali, she is the real form of all vidyas (goddesses) who give siddhi (success). Because Shiva and Shakti are one, they must be worshipped together.

Shiva then outlines the chief mantras of Dakshina Kalika. He then describes the dhyana (meditation form). A devotee should worship Kali, using vira bhava, as formidable, with rising swelling breasts full of milk, the colour of a thundercloud, dusky, roaring terribly, and having four arms. She carries a newly severed head, and a sword in her upper left and lower left hands. In her right, she shows the mudras dispelling fears and granting boons. Around her bloody throat, is a necklace made of 50 skulls which are the letters of the alphabet.

Two streams of blood trickle from her mouth. Around her are terrifying jackals which roar in the four directions. Her girdle is made of hands of corpses and she laughs. She is naked, with dishevelled hair, and bears a crescent moon as her diadem.

She is seated on the corpse form of Mahadeva, where she has intercourse with Mahakala in the viparita position. Her eyes roll with liquor, her smiling face is like a lotus and she is very terrifying Maharaudri who gives all bliss.

The chapter then describes vira sadhana at night in the cremation ground. A sadhaka should first worship mentally and then may do the outer form of puja. He should also worship Mahakala, whose dhyana is given as follows: Of a smoky colour, with matted locks, three eyes, united with Shakti, naked, of terrifying form, his effulgence equal to a sapphire unguent. He is both with qualities and without qualities.

Then a sadhaka should worship the 15 Kali Nityas in the five triangles and in the eight petals of the Kaliyantra should worship Brahmi, Narayani, Kaumari, Maheshvari, Aparajita, Chamunda, Varahi and Narahasimhika, from the east first. In the four doors of the yantra are Asitanga and the other seven Bhairavas who are Ruruchanda, Krodha, Bhishana, Unmatta, Kapali and Samharaka. They are given worship in pairs, from the east in order.

In the 10 directions, a sadhaka should worship the dikpalas (lords of the directions). After this worship, the practitioner should meditate on her in her form as Kulluka, using a mantra with five syllables situated in the different parts of the body. Shiva says Kulluka is Tara as Mahanilasarasvati. Following this, one should recite the mantra 108 times, worship Mahakala again with Lalita and recite the armour (kavacha) and the hymn (stava).

Chapter three speaks of the kavacha (armour) of Dakshina Kali. This, unlike other Kali kavachas translated in *Magic of Kali*, is brief and starts: 'Siddhakali, protect my head, Dakshina protect my forehead! Kali protect my mouth always, Kapali, protect my eyes. Kulla shield my cheeks always and Kurukullika protect my mouth. Virodhini protect the adhara (?) and Viprachitta the lips.

'Ugra, protect my ears always and Ugraprabha my nostrils. Dipta shield my throat and Nila be protective of my lower throat.

'Ghana protect my chest and Matra always protect my diaphragm. Mudra always protect the navel and Mita shield my lingam always.' The kavacha goes on to use the 22 letters of the Kali mantra to protect other parts of the body. Shiva then gives a hymn to Kali called the Kalika Stotra. This is essentially an extended meditation, similar to the above.

The main subject matter of chapter four concerns purashcharana, the preparatory acts a sadhika or sadhaka must perform before she or he becomes competent to recite the mantra. Shiva first gives a set of asanas or postures and then says there are 72,000 nadis in the body. The chief nadis for prana (bioenergy or vital breath) are 10 and of these the most important are the Ida, the Pingala and the Sushumna nadis. Within the last is the Chitrini nadi. The three nadis are the Moon, Sun and Fire devatas while Sushumna is of the nature of Sun and Moon conjoined.

Shiva then describes the 10 vayus, which are in sets of two. When a yogi unites that which is above and that which is below, he unites Sun and Moon, realises Om and is one with Hamsa. Hamsa, the Tantra explains, relates to the breath. The letter Ha is exhalation and the letter Sa inhalation. A living being (jiva) recites this supreme mantra known as the Ajapa mantra (that which is not recited) 21,600 times day and night.

The ajapa mantra is called the gayatri of yogis and gives liberation. This, says the Tantra, is the secret preparatory act before a mantra can become siddha. There follows a meditation on Kali where she is conceived of as light extending from the feet to the top of the head. A yogi or yogini should offer fruit, flower, scent, clothes, gems mentally to Kali again and again. This, explains Shiva, is the preparatory act of the northern amnaya of Kali Kula.

Shiva then allocates different forms of the Devi to the different directions and describes the purascharanas. A pashu, established in the southern amnaya should use the 22 syllable mantra and recite it two lakhs (100,000), half in the day and half at night. Every tenth time, the worshipper must give sacrifice.

Vira purashcharana is different. The sadhaka and his shakti should be naked in the cremation ground. The shakti should have Kamakala written on her forehead and in the centre of that should be the Devi mantra. The mantra should be recited 100,000 times and every tenth recitation should consist of an oblation of alcohol into fire. If a sadhaka does not have a shakti, he can worship her mentally.

Without doing preparatory acts, there is no entitlement to worship. Puja done without the preparatory acts makes black magic out of a person's recitation and sacrifice. One is also to give gifts to the guru and to his shakti and his relatives. Success in mantra cannot be achieved without supreme devotion.

Chapter five speaks of the Rajani ("the coloured or dark female"), which here seems to mean the shakti of a sadhaka. She should be free of shame, free of the opposites (dvandva), devoted to Shiva, pure (satva-gata) and by her own will (svecchaya) takes the viparita posture in intercourse. A sadhaka may also meditate on her mentally, as a mass of light in the brow. In this supreme form she sheds nectar. She should also be meditated upon as Gayatri in the form of exhalation and inhalation. This, says Shiva, is the Brahma Gayatri of yogis.

To obtain success, a yogi must reject greed, lust and envy. If a yogi does puja prompted by these, he goes to the Raurava Hell and becomes miserable. He is to reject the idea of difference and then achieves liberation. No-one should worship Kali if hungry or thirsty. 'After eating and drinking, one should worship the auspicious Kalika.'

Unless one is a vira or a divya, one should not worship Kalika. That brings sorrow 'at every step' and a person goes to Naraka Hell. One should not worship Kalika if lazy, as that will bring an individual to the level of a pashu (beast). The Kalika darshana is the lata darshana, that is the revealed doctrine into which sexual intercourse, likened to the twining of a vine, enters. It should be performed in an empty place, in a cremation ground, at a river-bank, on a mountain. There, one should worship Shakti. Without a guru, one should not perform ritual intercourse, which leads to hell, destruction and poverty

In vira sadhana of Kalika one should use meat, wine, flesh, fish and maithuna (the five m's). The text describes forms of the Devi who are worshipped in this fashion.

Chapter six speaks of the siddhi which ensues from worship of the Rajani, the chief of which is liberation whilst living. Shiva says that this knowledge, which destroys Samsara, should never be revealed. He then describes vira sadhana. The union of female and male is the supreme essence and is the worship of Kalika. It gives siddhi and is hard to obtain even for the gods.

In chapter seven, Shri Devi asks about abhisheka. Shiva says there are two types, that which is done in the vaidika way and also knowledge (jnana) abhisheka, which is hidden in all the tantras. He says a tantrik should do Kula-abhisheka, which creates peace, all that is good, dispels ailments, gives wealth, destroys great sins and the like. It gives the fruit of all bathing places (tirtha).

It is to be obtained from the guru. Shiva says that the devatas are not satisfied unless there is bliss coming from the worship of Kali and the five m's. Without Kulachara, it is impossible to be successful in the Kali mantra. Without this type of abhishekha, all puja turns into black magic and an individual goes to Naraka Hell or worse.

One must bow to the true guru, to deva and devi, do guru puja and then perform the abhishekha at the root of a bilva tree, at the junction of three paths, in the ancestral ground, in a deserted place and in other favoured Kaula spots.

Chapter eight opens with Shiva talking of arghya and the establishment of a pot to do the purashcharana. He gives the mantras associated with the worship. The chapter speaks of the Mahapuja, or great worship, and goes on to list at great length the different devatas of the tradition connected with the abhisheka. This worship gives success to a sadhaka.

In chapter nine, Devi asks how a person becomes successful in the mantras. Shiva describes the initiation of a Kula Shakti. After drawing a Kamakala yantra, the sadhaka should whisper the root mantra in her left ear. The initiated shakti sits on the left of the sadhaka, wearing red clothes, smeared with various scents and adorned with different jewels. The mantra should be drawn on her forehead. By worshipping this shakti in the Kulalula rite, devis from everywhere are attracted to the chakra. This rite produces nirvana for gods and for men. Intercourse with an initiated shakti brings success, provided the participants are initiated by the guru, otherwise the sadhaka is cast into the Naraka underworld.

Devi says she still is unsure about the different Shaktis and asks Shiva to explain further. Shiva says he will speak specifically about the Kula sadhana. A person should not do Kula sadhana without an initiated Vira Shakti.

He speaks of five chakras where these shaktis may be worshipped, which are the Raja chakra, the Maha chakra, the Deva chakra, the Vira chakra and the Pashu chakra. Brahmacharis and Grihasthas (householders) can worship in these five chakras. He speaks of various substances used in the chakras including svayambhu, kunda, gola and udbhava flowers, which are Kaula tantrik code-terms for menstrual blood and also gives days of the waxing and waning Moon which bring success in the particular rites.

The goddess asks who are the five mothers worshipped in the rite. Shiva explains they must be initiated women. Without worship of svayambhu, gola, kunda and udbhava flowers, the rites are useless and bring harm to sadhakas.

The tenth chapter is a discussion about the different chakras in the Kaula tradition, here meaning assembies of the Kaula folk. Shiva says that Vira or heroic sadhana may only be accomplished with an initiated Shakti. Chakras are of five types: Raja, the Mahachakra, the Devachakra, the Virachakra and the Pashuchakra. Both Brahmacharis and Grihastha (householders) can worship in all five.

Shiva gives details of the ritual accessories (upacharas) employed in these as well as the best times for creating them. The best times are the eighth and fourteenth days of the waning Moon on a Tuesday or on the fourth and seventh days of the waxing Moon on a Thursday. 64,000 forms of the Devi dwell in the different chakras. The Vira Chakra should take place on an eighth or fourteenth day of the dark fortnight in the ancestral grounds, that is the cremation ground. Shiva completes this chapter by describing the five maidens (Kanyas) and their worship.

Devi asks Shiva about sadhana of the yoni in chapter eleven. First, Shiva describes the characteristics of the sadhaka, then moves on to the sadhika. The male should be free from duality, ego-less, generous, fearless, pure, devoted to his gurudeva, peaceful and devoid of shame and greed. He should wear red clothes and red gems. The sadhika (female worshipper) should have similar qualities and when having intercourse should, by her own will (svecchaya) assume the inverse sexual position (viparita). She should be initiated in the tradition.

The mantra should be recited 108 times and the Devi worshipped internally. The Kulachakra should be drawn using vajra flowers and the preparatory act completed by reciting the mantra 108 times. The mantra Hrim should be drawn on the forehead of the Shakti. Shiva says that without puja of Kamakhya, it is impossible to be successful in the mantra. More details are given in chapter twelve.

In chapter thirteen, the Devi asks Shiva about the Vidyas (female mantras) giving siddhi (success). Shiva relates the different goddesses to the types of Shakti. Tara is the Chandali, Shri Vidya the Brahmani, Cchinnamasta the Kapali. This chapter also gives results from worshipping Devi for a given period of time. Doing Kaula puja according to the Mahachinachara rule and worshipping Kamakhya in a cremation ground gives sovereignty.

The quite long chapter fourteen opens with Devi asking Shiva about the Veshyas. This word, literally, means whore but is applied in this Tantra to initiated Shaktis and to Devis. Shiva enumerates seven, the Gupta Veshya, the Mahaveshya, the Kulaveshya, the Mahodaya, the Rajaveshya, the Devaveshya and the Brahmaveshya.

The Guptaveshya is without shame, with her eyes rolling with lust. The Mahodaya, by her own will, takes the viparita position. The Kulaveshya is the spouse of the Kula. The

Mahaveshya is a digambari by her own will, that is, she goes naked. Then follows a eulogy of the viparita position. If mantra is recited when in intercourse with the Shakti, she is Kali and gives mantra-siddhi and nirvana. When semen is emitted during the rite, a sadhaka becomes like Mahakala while the sadhika becomes like Dakshina Kalika. Only through the Kaula rite does a human being become enlightened.

The fifteenth chapter deals with the panchamakara, known as the five M's, and used in vamachara rites. These are madya (wine), mamsa (flesh), mina (fish), mudra (bean) and maithuna (sexual intercourse). The sadhaka, at night, sits with his Shakti to his left, doing the various types of nyasa first. The chapter gives the rules of puja and towards the end enumerates the ten Mahavidyas and the other Siddha Vidyas. These are given as Kali, Tara, Cchinna, Matangi, Bhuvaneshvari, Annapurna, Nitya, Durga, Mahishamardini, Tvarita, Tripuraputa, Bhairavi, Bagala, Dhumavati, Kamala, Saraswati, Jayadurga, and Tripurasundari. For these 18 Mahavidyas, there is no need for purification, nor of considering day, tithi, nakshatra, yoga or karana. Thus ends the *Niruttara Tantra*.

Yogini Tantra

The Yogini Tantra is a voluminous work held in high regard by practitioners of Vamachara. In a total of 28 chapters divided into two parts, it outlines every topic familiar to the Kaula and Vama traditions. What follows is an abstract of the first nine patalas or chapters.

First Patala

This opens with a familiar tantrik scene on Mount Kailash where Shiva is addressed by Parvati. She says she has heard exposition of tantras before on Shri Shaila mountain, in Varanasi, in Kamarupa and in Nepal. Now she wants to hear more from Shiva, the world guru. In answer, Shiva says he will declare the great Yogini Tantra, the giver of both wealth and liberation. It is to be concealed and is unknown to all the devatas, to the asuras, to the yakshas and others but he will declare it out of love for Parvati.

He starts by eulogising the goddess as the cosmic mother (Vishvamata), dark as a thunderstorm, wearing a garland and waist-band of skulls, with dishevelled hair, completely naked (digambaram). She has a rolling tongue, makes a terrifying roar, three reddened eyes, and has a wide open mouth. She wears a moon digit on her forehead, has the corpses of two boys as her earrings, and is adorned with various gems, which are of the brightness of the Sun and the Moon. Laughing loudly, she has two streams of blood pouring from her mouth, while her throat is red with blood. In her four arms she holds cleaver, head, and makes mudras dispelling fears and granting boons. She, the supreme Nitya, is seated in reverse (viparita) intercourse with Mahakala upon the corpse of Shiva. The whole scene is set in the cremation ground.

After this detailed dhyana of Kali, Shiva begins to outline the Tantra, declaring that he is Parvati's slave. He starts with the characteristics of the guru, who he describes as the root of all shastra, the root of this world and the very self of Parabrahma and the essence of Shiva. The guru can save a disciple where even gods and goddesses cannot intercede. The guru's family is to be considered as identical with the guru. There follows a dhyana of guru in the palace of wish-fulfilling gems on Mount Kailash, surrounded by hosts of Bhairavas. The palace is surrounded by the seven oceans. The guru is one with Mahakala Adi Natha and knows all mantras, whether they be Shakta, Vaishnava or Ganapatya. The greatness of the guru is hymned in all the shastras.

Second Patala

Devi asks Shiva to speak of Kali and Tarini. Shiva says that Kalika is the greatest of the great vidyas, supreme and giving nirvana and liberation to people. Her disciples are Brahma, Vishnu and himself. If a sadhaka recites the Kali mantra, he becomes her son. Kali, Tara and Cchinna are the mahavidyas. One successful in Kali becomes similarly successful in the others. Shiva begins to speak of initiation. He says that the rosary to be used in the puja should be made of human skullbone for long-lasting success. A sadhaka or sadhika may also use crystal or ruby rosaries. A full rosary should have 108 beads. The meru, or bead to mark the beginning and the end of the mala, should be made of a king's tooth. Shiva proceeds to outline the number of times the mantra should be recited holding the rosary and the way the fingers should count. He speaks of the nature of other rosaries including pearl, tulsi (basil) when worshipping Vishnu, ivory for Ganesha, and rudraksha or red sandalwood for Tripura.

Dhattura growing in a cremation ground is used for Dhumavati. He then describes ritual accessories to be used in the puja and the times in bright and dark fortnights of the moon which are favourable and unfavourable as well as other restraints due to time as well as suitable places for the rite.

Third Patala

Devi asks how catastrophes including war and fever can be warded off. In reply, Shiva recites a kavacha or armour which can be used to protect against malefic influence. It is not to be revealed lightly. He then speaks of a way to subjugate the world (jagadvashyakara). Sage Narada also asked Shiva to speak of this of old. Shiva says that when she is imagined as a naked Devi, Kali is the deludress of the world. He then gives the Trailokya Mohana Kavacha (armour bewildering the three worlds). Kalabhairava is the rishi of the mantra, anushtubh is the metre, Shmashana Kali is the devata.

After giving the armour, Shiva describes how to make it. It should be written on bhurja (birch) bark and worn round the person. It should be written on the eighth day of the bright fortnight and placed inside a golden container. Wearing it on different parts of the body gives different results. On the head, it destroys disease. On the right shoulder, it gives whatever is desired. Vishnu now chimes in and says Narada achieved the desires he wanted by employing this kavacha.

Fourth Patala

The Devi now wants to know of other prayogas to give dominion, knowledge and wealth. Shiva mentions the *Phetkarini Tantra* and the *Nila Tantra* as sources. One process is to draw a hexagon with the mantra of Tara within plus the sadhya (the object). Devi asks about the satkarmas, six magical acts. Shiva says these are pacifying, subduing, causing enmity, driving away, uprooting (uccatana) and causing death. He says there are six Shaktis appropriate to these acts. The Padmini is suitable for pacifying and Sankhini for subjugation. He then outlines the mantras appropriate to the six acts.

Fifth Patala

Shiva describes a great sadhana in the cremation ground, involving the fifteen Kali Nityas. This sadhana can also be performed in a desert, by the side of a river, on a mountain, at a crossroads, at the root of a Bilva tree, at a place where there is a single lingam, at a place where there are no people as well as in the cremation ground.

Sixth Patala

Devi asks about the different classes of sadhaka. Shiva says they are divided into the divya (divine), vira (heroic) and pashu (herdlike) categories. The meditation for the divya should be concealed, Shiva speaking of vira meditation. He says a vira should meditate on the three bindus in the form of a 16 year old woman. The first is as bright as 10,000,000 dawn suns, extending from the head to the breasts.

The second extends from the breasts to the hips and the third from the yoni to the feet. This is the Kamakala form, the very essence of Brahma, Vishnu and Shiva. The vira and the divya may employ madya, mamsa, matsya, mudra and maithuna (the five m's) in their worship. According to Shiva, the rishis, the vasus, the daityas all became great through this puja. Shiva speaks of this worship for the four Hindu divisions (varnas) and also for the Avadhuta. Much of this material is repeated in the *Yoni Tantra*. He says the great nectar flows from Kundalini when it has risen to the top of the head. This is the great wine. She is the supreme Shakti within the body.

Seventh Patala

Devi asks Shiva to tell her of Devi Svapnavati (she who

moves in sleep). He gives the mantra. Shiva says: 'This Svapnavati Vidya is very hard to obtain in the three worlds. It is the cause of great miracles, declared by Mahakala.' It should be recited 108 times then Svapnavati visits in sleep always. A sadhaka who masters the mantra sees everything in his dreams he wishes.

The god then speaks of the Mritasanjivani vidya. This appears to give the power of bringing back the dead to life. He describes other vidyas including Madhumati and the Trilokyakarshi vidya. This attracts whatever a sadhaka desires in the three worlds. Maidens will cross oceans and mountain ranges to get to him. Shiva goes on to give vidyas of Padmavati, the Vashikarani vidya. He then returns to the topic of Svapnavati. This appears to involve awakening while in the dream state (lucid dreaming?). The mantras should be concealed and given only to the devoted, the unstained otherwise hosts of Dakinis consume a person.

Eighth Patala

Shiva speaks of the Yoginis. They look terrifying, with blazing eyes and 50 lakhs of faces. The daitya Ghora then recites a hymn to Devi, celebrating her victory over the Daityas. Shiva chimes in, praising her greatness in battle. Towards the end of the patala, Shiva gives a meditation image of Shakti as Kali.

Ninth Patala

Shiva starts this lengthy chapter by speaking of the Devi as the Brahmanda, the macrocosm. In this guise, she has an immense form, with millions upon millions of arms and heads. She is the sum of everything, containing puranas, vedas, smriti and vedas. As such she is of the brilliance of millions upon millions of suns and moons and fires, consisting of all knowledge, all paths,. all dharma, all bliss, all shastra, all veda and all worlds, in short, everything. Then follows a meditation on Shakti as being present in the different parts of the body. Shiva closes by saying that Kali is the form of consciousness (citrupa), the impartite absolute.

Tantra of Kulachudamani

Kulachudamani Tantra is a nigama, meaning that instead of Devi asking questions answered by Lord Shiva (agama), he asks questions answered by Devi, the goddess. In this Tantra, the cult goddess is Mahishamardini, a Devi with some similarities to Durga.

In seven short chapters, Devi expounds the essence of her worship, sometimes in beautiful and nearly always in colourful language. But the uncanny side of Kaula and Kali worship is dwelt on in great detail, with references to siddhi, including a mysterious process where the tantrik adept leaves his body at night, apparently so he can engage in sexual intercourse with Shaktis.

These uncanny elements may well have code meanings and be intended to throw the unwary off the scent. See, for example, the Jnana Karika, which gives an entirely different slant to crossroads, Kula trees, Kula wine and the like.

Animal sacrifice also appears to have a place in this Tantra, as well as gruesome magical matters, including using the bones of a dead black cat to make a magical powder. Please don't try this at home.

To read the abstract of the text provided in Sir John Woodroffe's *Tantrik Text* series (here) you would have little idea of this.

When, under the pseudonym of Arthur Avalon, he produced this series, at the beginning of the 20th century, the fact that an Englishman and a High Court judge would interest himself in even orthodox Hinduism no doubt caused eyebrows to rise and lips to be pursed amongst his peers.

As well as including that introduction, we also here provide our translation, first published in *Azoth* magazine in the early 1980s.

Although far from being completely happy with this rendition, it does give a flavour of the content which you would be unlikely to get if you relied just on the *Tantrik Texts* abstract. Chapter seven remains to be translated.

The siddhis-or magical powers-play a large part in this text. The main tantriki rites are called the six acts (shatkarma) of pacifying, subjugating, paralysing, obstructing, driving away, and death-dealing.

But the Kulachudamani includes others such as Parapurapraveshana, which is the power of reviving a corpse; Anjana, which lets a sedhaka see through solid walls; Khadga which gives invulnerability to swords; Khecari, which gives the power of flying and Paduka siddhi, magical sandals which take you great distances, rather like seven league boots.

Certainly, the importance of having a suitable Shakti forms the essence of the instructions Devi gives to Shiva. We see this emphasis over and over again, throughout the Tantra.

Devi here takes the form of Mahishamardini, more popularly known as Durga, who destroyed the two arch-demons Sumbha and Nisumbha in an epic battle between the goddess and the throng of demons. It was at this time, according to legend, that Durga created Kali, by emanating her out of her third eye.

We learn more of Durga's legends and myths from the *Kalikapurana*, an influential source in Kaula Tantra. The Devi, Mahamaya, appeared as Bhadra Kali-identical with Mahishamardini-according to the same text, in order to slay the demon Mahisha. He had fallen into a deep sleep on a mountain and had a terrible dream in which Bhadra Kali cut asunder his head with her sword and drank his blood.

The demon started to worship Bhadra Kali and when Mahamaya appeared to him again in a later age to slaughter him again, she asked a boon of her. Devi replied that he could have his boon, and he asked her for the favour that he would never leave the service of her feet again. Devi replied that his boon was granted.

"When you have been killed by me in the fight, O demon Mahisha, you shall never leave my feet, there is no doubt about it. In every place where worship of me takes place, there (will be worship) of you; as regards your body, O Danava, it is to be worshipped and meditated upon at the same time."

First Patala

Sri Bhairava said: Innumerable are Tripura (tantras), innumerable those of Kalika, countless those of Vagishvari, numberless the beautiful Kulakulas known as Matangini, Purna, Vimala, Candanayika, Tripura-Ekajata, Durga and Kulasundari.

Numberless the Vaishnava, Ganapatya, Saura, Shaivite, and the different doctrines of Shankara.

The highest of the doctrines are the 64 tantras: the Mahasarasvata, Yoginijala, Shambara, Tattvashambara, the eight Bhairavas, the eight Bahurupas, the Jnana, the eight Yamalas, the Tantrajnana, the Vasuki, the Mahasammohana, the Mahasuksma, the Vahana and the Vahana Uttaram, the Hrdbheda, the Guhyatattva, the Kamika, the Kalapaka, the Kubjika Mata, the Maya Uttara, the Vina, the Trodala Uttara, the Pancamrta, the Rupabheda, the Bhutadamara, the Kulasara, the Kullodisa, the Visvatmaka, the Sarvajnatmaka, the Mahapitrmata, the Mahalakhmimata, the Siddhi Yogesvari Mata, the Kurupika Mata, the Rupikamata, the Sarvaviramata, the Vimalamata, the tantras of east, south, west and north, the Niruttara, the Vaisesikajnanatantra, the Shivabali, the Arunesa, the Mohanesa and the Visuddhesvara.

O Lovely Hipped One, now speak to me of their essence if you have love for me.

Devi said: Listen Deva, supremely blissful quintessence, the Lord of Kula, to the very essence of knowledge of the ocean of Kula Tantra, concealed by my Maya.

I am Great Nature, consciousness, bliss, the quintessence, devotedly praised. Where I am, there are no Brahma, Hara, Shambhu or other devas, nor is there creation, maintenance or dissolution. Where I am, there is no attachment, happiness, sadness, liberation, goodness, faith, atheism, guru or disciple.

When I, desiring creation, cover myself with my Maya and become triple, becoming ecstatic in my wanton love play, I am Vikarini, giving rise to the various things.

The five elements and the 108 lingams come into being, while Brahma and the other devas, the three worlds, Bhur-

Bhuvah-Svah spontaneously come into manifestation. By mutual differences of Shiva and Shakti, the (three) gunas originate.

All things, such as Brahma and so forth, are my parts, born from my being. Dividing and blending, the various tantras, mantras and kulas come into being. After withdrawing the fivefold universe, I, Lalita, become of the nature of nirvana. Once more, men, great nature, egoism, the five elements, sattvas, rajas and tamas become manifested. This universe of parts appears and is then dissolved.

O All-Knowing One, if I am known, what need is there for revealed scriptures and sadhana? If I am unknown, what use for puja and revealed text? I am the essence of creation, manifested as woman, intoxicated with sexual desire, in order to know you as guru, you with whom I am one. Even given this, Mahadeva, my true nature still remains secret.

Devi said: Listen son, to the very plain exposition of the teachings of utmost bliss. I speak of the method relating to the yoga of liberation. This is the one essence of all tantras, worshipped by all devas, giving every sort of knowledge, secret, giving a clear idea of the essence of enlightenment, free of good or evil, giver of both enjoyment and liberation, consisting of all paths. Dear son, it deludes even the wise!

It is of various and numerous meanings and is the goal and refuge of all pure disciples. It is known as the best path according to all the paths, yet is reviled by all doctrines. It should be known only through the best of teachers. It should be protected and hidden in the heart very carefully, as I did not reveal it to Vishnu, Dhatra or Ganapati.

Dear son, whoever is unaware of this Tantra is incompetent. Dearest, I speak of the pure knowledge of Kulachara.

Arising at dawn, a sadhaka should bow to a Kula tree. After meditating on the Kulas from the Muladhara to the one thousand petalled lotus, he should meditate on the gurus.

Then he should meditate on and worship the Kulagurus Prahladanandanatha, Sanakanandanatha, Kumaranandanatha, Vasistanandanatha, Krodhanandanatha, Sukhanandanatha,

Jnananandanatha, and Bodhanandanatha as being intoxicacted by the nectar of the absolute, whose blissful hearts show in their eyes, their darkness cut and crushed by their having embraced Kulashastra, merciful to Kula disciples, complete, compassionate and effulgent, giving boons and dispelling fear, knowing the essence of all Kulatantras.

After bowing to the Kulagurus, one should give worship to the Kulamatrikas, fashioning a Kula place and bathing oneself in all the holy waters (tirthas).

Dear son, a Kulaguru is an accomplished being and said to be the vehicle of happiness. Dear son, conceal this very secret marvellous doctrine from the sight of pashus.

They who should reject the Kulanatha, who alone is a Shakta, served by the Kulas, for them initiation and sadhana is black magic. Because of this, one should by every effort resort to a Kulina guru. It is said a Kulina is competent in all vidyas and is able to initiate in all mantras.

Second Patala

Devi said: Dear son, now I will speak of bathing, the vehicle of Kula happiness. I have various forms (coloured) black, red, yellow and blue. Whichever pupil on the path of Kula goes to bathe, attains my form. Everything, heaven and hell, originates on earth. After sipping water, strew the Kula place with grass and Kula flowers and place durva grass, sesame oil and water in the Kula vessel.

After satisfying the Kuladeva, bathe. At first performing resolution, then draw the Kula Cakra on the (surface) of the water. Bow to the Kula trees, and using the Kula mudra called Ankusha, the Kula should invoke the Kula tirthas (into the water). After drinking the water three times, bathe body three times. Dedicate the offering to the deva of the Kula tree three times.

Using the Kula water, oblate the devas, the ancestors and the rishis. Again, after meditating on the Kulas, offer water to the Kula devas again. In the Bhairavi Tantra there are verses relating to this knowledge: Bhairavaya devaya. Creation came from Bhairava. Offer to Bhairava, pronouncing this mantra. After giving a suitable offering, meditate on the being of Bhairava-Bhairavi, offering the remainder.

Deva, by meditating in this manner, I bestow grace, whether the rite be that of ancestors, Shakti, offering, bathing or limb puja, there is no doubt about this. After satisfying (Devi), offer the remainder to the people pervading the world who have this thing. Then rise, don the Kula robe, and envelop oneself in Kula.

Making a forehead mark of the Kula type, sip water again. Pay respect to the Kula pitha and do worship of the Kuladeva.

After satisfying the guardian of the door by suitable song, dance, speech and so forth, one should collect the Kula elements and purify the Kula seat. The Kula seat duly prepared, then strew the area with suitable pleasant things. Sitting on the Kula seat, and binding the hair, do the ritual of Guru puja.

A person should purify himself, the area of ritual work, and his own body. Sipping the offering, the wise person should then worship the Kula ishtadevata. Do the puja with initiates, with adepts, with young maidens, with Kula people, and with those devoted to devata and guru.

Various kinds of flowers and different sorts of scents should be present and one should don clothes scented with camphor and incense, smeared with scented powder. Offer tambula and various other pleasant substances, giving incense and fire first.

A Kaulika should wear all kinds of jewels and gems, and, reciting the root vidya, should sprinkle the place with water. All the substances should be on the right, while the offering (arghya) should be on the left. The Kula substances should be to the west of the devata.

Making a yantra using different menstrual flowers such as svayambhu and different red materials like rocana, lac, kumkuma, and red sandalwood, one should do the puja, afterwards offering recitation. After reciting the Shakti mantra and praising her, then do the dismissal. Circle, and then prostrate yourself in front of the young woman there present.

After previously offering the essence of the Kula nectar to the guru, one should then eat food.

One should worship the young woman and she should worship you. Conceal the design of the yantra in the secret place of the 1,000 petal lotus. Only impart this to a Kulina and never to atheists, fools, pashus or brahmanas, otherwise one meets with death.

Folk having gone at night to the cremation ground or to a Kula house and placing in the centre of flowers and sweet scents the highest Kula thing, should, in the company of Kulas, within the Kulacakra, draw a Kulayantra containing the name of the object to be accomplished. After first writing one's own name, accomplish the sadhana following the rules of the Kulacara. The sadhaka should do the sadhana with his own and other Shaktis.

Dear son, now listen to the rules relating to unification with Parashakti. Embrace one's own partner, who should be very beautiful and very alluring. One should act as the guru to the Kula devotee, and should initiate her into the path of Kula.

(She should) show in her eyes the very blissful essence, be Kula born, be faithful and very wise, inwardly protective of the guru, with her mouth full of tambula.

You should worship her as if she were your own daughter. Then draw on her forehead a Shakti Cakra of three concentric circles, within this writing the Kamakala mantra. In the centre, using mantra, write the name of the object to be attained. Inside this, invoke Devi and after meditating on her, worship her.

Then pronounce the rishi, metre and root mantra into the ear of your daughter, three times in her left ear.

Son, now listen to the sexual embrace in Kula puja. A knower of Kula should worship she who is wanton and free from shame, doing the actions according to the guidance of the guru. After being initiated, prostrate yourself like a stick on the ground.

Say Save me! O Lord of Kula, who with your Padmini is on the path of Kula! May the shadow of your lotus feet fall on my head, O Princely One! After giving dakshina to the guru with eyes full of love and tambula in his mouth, accomplish whatever you want with your own Kula Shakti.

If, firstly, you do not do limb and avarana puja, then you are not a Kula. After first meditating on one's own guru as being above one, the very essence of Kula nectar, and after oblating that deva, then one may recite mantra.

Third Patala

Now I speak to you of worship at night in one's own Kula. The Shakti should be seated on your left hand side on a mattress, adorned with red clothes, bejewelled with gold, smeared with red scents, garlanded with flowers, perfumed, wearing bright things, very beautiful, wearing lovely clothes, with eyes like shy blossoms, slender, with large full breasts.

On her forehead, draw a beautiful yantra and in this write the object to be accomplished. Draw the same on her shoulders, arms, breasts and stomach. Her mouth should be filled with tambula and Kula substances. After doing recitation of the Kulakula mantra, one attains the desired for thing immediately.

She comes from a distance of 100 yojanas, across rivers and mountains, across 1,000 isles, free of restraint, with agitated eyes, shedding love juice profusely, trembling, the circle of her beautiful buttocks swaying, her heart full of love for the sadhaka, boldly coming ever closer, coming to sit with the sadhaka, moving like the devata. Attracted to him in this way, a sadhaka achieves success and becomes a Kaulika.

Unless she is initiated and young, how is it possible to accomplish Kulapuja? There can be no Kulapuja, dearest, unless she has previously obtained the Kula mantra. When other than young, dear son, it is as if she were one with the guru (?).

In her left ear, recite the mantra while sprinkling her (with Kula nectar). Mahadeva, now listen. I will expound this mantra to you. Aim Klim Sauh Tripurayai imam Shaktim pavitram mama Shakti kuru Svaha. This mantra has 26 syllables (?). O Deva, purify the Shakti using this mantra.

Brahmin-girl, warrior-girl, merchant-girl, slave-girl, Kulini, daughter of a barber, washer-girl, yogini; these are the eight girls.

Each woman is equivalent to a Kula maiden. Hold the Kula cakra at a crossroads, close to a river, at the root of a bilva tree, actually within the cremation ground, during a feast, in a palace or whatever, O Holder of the Trident!

Draw, dear son, a great yantra using powdered vermilion, and strew on it the names of the object you want to accomplish!

Use couch grass. Worship according to the rule and enter the Kula using the essence of Kula. Worship therein according to the due prescription and create the Kula using the essence of Kula!

Offering wine within the Kula area, devoted souls should then worship beautiful young women, duly initiated, wearing (silken) robes, garlands and so forth, giving them food, good milk and all the rest.

To start, give (these maidens) food you have cooked yourself (dear son). (Examples follow): different sorts of cake, curd, milk, ghee, buttertmilk, candies. Offer different side-dishes flaboured [for example] with crushed saffron and essences of variegated sources (drawn from) the art of cookery.

Try offering jackfruit, polished cardamoms, washed lemons, pomegranates and different other pleasant fruits [all the while] smearing the maidens with a variety of scents and perfumes.

Try offering them sandalwood, musk, saffron, fresh green sprouts of pallava, borax, blossoms of the Lodhra tree, things from the water, items from the forest. Bring [your beautiful Shaktis] different jewels, and decorate them-in turn-with very precious jewels of different kinds.

Do the worship in a private place, and give offerings and also do purifications. Once [a sadhaka] has caused the Kula amrita to flow, he should bow in front of the Shaktis.

He should bow to each of the Shaktis, in turn, and should call out their names, starting with the Brahmini. Asking each to take a seat, a sadhaka should make sure each one has a seat.

Then he should give them offerings, water to drink, water for each [beautiful goddess] to wash her ,toes, honey-flavoured water, and water yet again.

If they are uninitiated, [the sadhaka] should say Hrim to each. O truly beauteous One, he should feed them in the centre of a pavilion using golden plates. Then he should recite the hymn.

Om hail to you Mother Devi!
Stainless soul, the essence of Brahma.
Through your compassion remove obstacles and bestow siddhi on me!
Maheshi, giver of blessings!
Devi, the form of supreme bliss!
Through your compassion &c.
Kaumari, who dallies with Kumara,
Lady of all Knowledge,
Through your compassion &c.
Vashnavi, carried on Garuda's wings,
The very self of Vishnu
Through your compassion &c.
Varahi Devi, giver of blessings,
Who lifted the earth on your tusks,
Through your compassion &c.
Devi, you are Aindri, worshipped by all the gods and Indra.
Through your compassion &c.
Chamunda, smeared with blood, dressed in a garland of severed heads,
You destroy fear!
Through your compassion &c.
Mahalakshmi Mahamaya, you destroy anxiety and sorrow.
Through your compassion &c.
Devi you are the goddess, father and mother both!
You take the place of our father and mother!
Although one, you are many, in the form of the cosmos!
Hail to you Devi, hail!

Fourth Patala

Dear son, my secret originates in simple practice. Those lacking this do not obtain success even in one hundred koti of births. Folk following the path of Kula and the Kulashastras are broad minded, from following the path of Vishnu, patient of insult, and always doing good to others.

One should go to the temple of a deva, or to a deserted place, free of people, an empty place, to a crossroads or to an island. There, one should recite the mantra and, having bowed, become one with divinity and free from sorrow.

Bow to Mahakali if you see a vulture, a she-jackal, a raven, an osprey, a hawk, a crow or a black cat, saying: "O Origin of all, greatly terrifying one, with dishevelled hair, fond of flesh offering, charming one of Kulachara, I bow to you, Shankara's beloved!"

If you should see a cremation ground or a corpse, circumambulate. Bowing to them, and reciting a mantra, a mantrin becomes happy: "O you with terrible fangs, cruel eyed one, roaring like a raging sow! Destroyer of life! O mother of sweet and terrifying sound, I bow to you, dweller in the cremation ground."

If you should see a red flower or red clothes-the essence of Tripura-prostrate yourself like a stick on the ground and recite the following mantra: "Tripura, destroyer of fear, coloured red as a bandhuka blossom! Supremely beautiful one, hail to you, giver of boons."

If you should see a dark blue flower, a king, a prince, elephant, horse, chariot, swords, blossoms, a vira, a buffalo, a Kuladeva, or an image of Mahishamardini-bow to Jayadurga to become free of obstacles. Say: "Jaya Devi! Support of the universe! Mother Tripura! Triple divinity!"

If you should see a wine jar, fish, meat or a beautiful woman, bow to Bhairavi Devi, saying this mantra: "O destructress of terrifying obstacles! Grace giver of the path of Kula! I bow to you, Boon-giver adorned with a garland of skulls! O red clothed one! One praised by all! All obstacle destroying Devi! I bow to you, the beloved of Hara."

Dear son, if a person sees this things without bowing, the Shakti mantra does not give success.

I all of this I am the most important part, beloved of the Kula folk. All the Dakinis are ıny parts. Listen Bhairava! One who has gained success in my simple yoga cannot be harmed by a Dakini. My devotees abound in wealth and cannot be conquered by Vatukas or Bhairavas.

Whichever Kaula is seen by a young girl or woman, whether he be in village, city, festival, or at the crossroads, causes her to be filled with longing, her heart aching, her eyes darting glances, like a line of bees mad for honey falls on a lotus flower, greedy for nectar, like a female partridge for a cloud, like a cow for her recently born calf, like a female gazelle eager for young shoots of grass, like jackals for flesh, like a person tortured by thirst who sees water, like a dvamsi (?) at the sight of a lotus fibre, or like an ant greedy for honey.

The sight of such a Kaula, enveloped by the Kulas, causes her lower garment to slip, she becomes mad with lust, and of unsteady appearance.

Seeing her on a couch, her breasts and vagina exposed, one should fall to her feet, and, rising, fall again. One should impart the oral lore to an alluring female companion-in her feet resides the secret of the act of love. One attracts such female companions, with beautiful hips and beautiful breasts, like a moon to the Kaula, free from greed or modesty, devoted, patient of heart, sensuous, very inner of spirt.

In such a happy Duti, curiosity may suddenly arise, she asking "Dear son, what is to be done or not to be done? Speak!" One should perform sacrifice to the indwelling Maya and offer the remainder to the Shakti. After this, one should excite her and then perform the act of love.

On a Tuesday, in the cremation ground, smeared with Kula vermilion, using Kula wood, one should draw a yantra. In the petals write the Canda Mantra, Sphrem Sphrem Kiti Kiti twice, and then the ninefold mantra of Mahishamardini. Outside this, write the mantras of Jayadurga and Shmashana Bhairavi.

After writing them, worship Bhadrakali at night, meditating on Kamakhya, the essence of Kamakala.

The Kulakaulika, naked, with dishevelled hair, should meditate on the formidable Kali, with her terrifying fangs and appearance, Digambari, with her garlands of human arms, seated on a corpse in Virasana, in sexual union with Mahakala, her ears adorned with bone ornaments, blood trickling from her mouth, roaring terrifyingly, wearing a garland of skulls, her large and swelling breasts smeared with blood, intoxicated with wine, trembling, holding in her left hand a sword, and in her right hand a human skull, dispelling fear and granting boons, her face terrifying, her tongue rolling wildly, her left ear adorned with a raven's feather, her jackal servants roaring loudly like the end of time, she herself laughing terribly and pitilessly, surrounded by hordes of fearsome Bhairavas, treading on human skeletons, wholly occupied with the sounds of victorious battle, the supreme one, served by numberless hosts of powerful demons.

After meditating on Kalika, the lord of Kula should then worship her. Unless one enters the other city, Kulasiddhi cannot be achieved. Because this Devi gives all success as soon as she is remembered, she is hymned in the three worlds as Dakshina.

O Bhairava, by reciting her mantra 108 times, one can achieve whatever object is wished for. After establishing oneself at the crossroads and meditating on the Devi in your heart, one should enter the city adorned with the most beautiful sorts of jewels. After meditating on Devi in the four directions, bow to the Kulaguru and, holding the name of the object of siddhi in your left hand, pronounce the mantra.

By smearing the eyes with anjana, one may shatter iron locks barring doors, becoming able to enter either stable, warrior's house, Kalika temple, treasury or sacred place, and may have sexual union according to will even 100 times. After meditating on Svapnavati Devi, one should enter the pavilion of Kaama.

Do puja with a yantra, writing the appropriate mantra on it, and reciting it at Devikuta, Oddiyana, Kamarupa, Tata (?)

or at Jalandhara or Purna(giri) on pure ground. Establish cakras in these places and, worshipping Devi, bow and recite her mantra eight, ten, 100 or 1,000 times. Reciting and offering at such a pitha, one gains the wealth of a treasury. On a pure spot, establish the siddha seat, preparing the protective pedestal and bowing to the pitha in a pleasant way. Say: "Come, O great one of the form of the vagina! Siddhayoni! Give that which is desired! I will perform Kulapuja with appropriate ritual accessories! Yield to me!"

Becoming like her son, her feet on your head, she yields whatever is wiṣhed for. Repeatedly offer her Kula flowers, scent and food. Dear son, prepare everything, and after offering and cooking for her, give grain, rice, wine, fish, flesh, ghee, honey and the other things which bestow success. A sadhaka should install her in a jar, and then worship the supreme. After meditating on the ishtadevata, the possessor of the path of Kula should feed her.

Dividing a piece of fruit in half, give one to the Kula Shakti, take the other one yourself, and then eat. If one does not have a young woman as a Shakti, perform the dismissal using water. After performing the pitha puja, rub out the yantra, offering then to the ancestors of the place.

Fifth Patala

Shri Devi said: Deva, a sadhaka travelling in dream can enter Kamarupa, which is Kamakhya, the yoni pavilion. After drawing the ultimate cakra, surrounding it with Kula substances, a sadhaka should write round it the named object of desire. Making the place of Kaama, in the centre of the place of Kaama, fashioning it into a funnel-shaped vessel, by Kaama one should love Kaama, turning Kaama into Kaama.

After meditating thus, and reciting a mantra, gazing at the pedestal and so forth, a possessor of the correct rule should place the father-face into the mother-face. So, offering cloth, saffron or tambula twice, one may take Kakini by force (?).

Doing ritual circling and so on according to the due rule, one may then go forth. If a man should attempt to subjugate women of the circle or protected property, a sadhaka becomes a fallen sadhaka. An act of black magic causes the destruction of a Kula. If he does a bad deed, it kills him, no doubt. City shaking prompted by black magic causes him to be bound by a superior power. The puravasini can be awoken using the sleep awakening mantra.

O Shankara, a Kulina using this method who intends to steal creates obstacles for himself, there is no doubt of it. Bhutas, pretas, pishachas, rakshasas, serpents, kinnari, naga maidens, underworld maidens, fairies, bhairavas, vatukas, and Ganapa create obstacles for those entering the place where protected women sleep. It causes the death of one's children, creates delusion, disease, and uprooting.

They cause different obstacles such as poverty and great anxiety, and loss of grain to one who emits his semen into a protected woman. Then they destroy sadhakas. One should protect oneself carefully, pegging all the gaps with vajra, Shakti, staff, sword, noose and goad. To avoid obstructions, carefully worship the guardians of the directions.

O Deva, one should offer cake, plantain, sweetmeats, milk rice, food, crushed parched grain and jackfruit, giving to Vishnu the supreme food and to Ganesha the pithaka (?). A vira should offer sweetmeats, jackfruit, plantain and black goat flesh to the Kshetresha, and then recite the mantra.

After reciting, offer a clump of earth and sprinkle water to the ten directions. Just as a horse sprinkled in sacrifice, as Sakra to the gods, so obstacle worship is for Kula. In the sleeping place, in the north east, place a Kula conch.

Around it, make a square of twelve finger breadths. A sadhaka, after making this king of yantras, should worship there at night. At night, roaming about, at night, doing Kula puja, there is nothing which cannot be done. A sadhaka becomes a Kaulika.

At night, establishing Tribhuvaneshvari, and bowing, recite her mantra. Bathing in the morning, and bowing to guru, devas, ancestors and rishis, offering oblation to the Shakti, worship in a devoted way.

O Bhairava, by serving a young woman who resembles a prostitute, one gains wealth, becomes all protected, beloved of all, and able to enslave. She bestows her grace. Kula sadhakas will know the different meditations by appeal to the Kulachudamani, previously spoken of. After offering in a gold, copper or a Kula vessel, and drawing the nijayantra or Kulayantra or the Shriyantra or the Gandharvayantra, made of various things, and with in the centre the name of the Kula target, strewn with nija names and the Kamakala bija, all encircled with the nijamantra, the best of sadhàkas should worship the essence of Kulamnaya.

With Kula puja and the like, using lingas, one may attain the highest core of Vishnu, saying Jaya Vishnu, Hare Brahma and so forth, offering various things, and doing the Kula puja in a forest or near a lake.

Using the previously declared rule, one may accomplish Kula agitation, etc. One should bring the Kula born Devi at night to a deserted garden, house or temple, and initiate her using the root mantra. Then, using the rule previously spoken of, one may achieve Kula agitation. Both should recite the root mantra which gives siddhi.

Of all pithas, the supreme pitha is Kamarupa, the great giver of results. O Maheshvara, whoever does puja there is accomplished.

Son, I live in this best of all pithas. Therefore the Kamakhya yoni mandala is spoken of 100 times. Mahadeva, what can be said of the fruit gained thereby? There dwell millions of Shaktis and Mahishamardini herself. This pitha is the image of the absolute, the hidden vehicle of all happiness.

Bhairava said: Deveshi, if I am truly your son, speak of the methods of attraction, you the cause of creation and dissolution.

Devi said: Dear Son, I will speak of the mahavidya, the supremely great attraction maker, through which puja method a man can attract even the devas. After reciting the Kali mantra of one, two or three types, one can attract the moveable, the fixed and everything else, according to will. This is revealed clearly, as if from the mouths of Brahma and Saraswati.

This Maha Kali vidya is said to be the ultimate secret of all secrets, causing sleep, wakefulness, delusion, confusion and bewilderment. One may go anywhere, whether in the night, the day, or the twilight. One should strew the name of the object to be accomplished (with the letters of the) bija mantra.

The guru should perform an act to enlighten the vira in this matter. Whatever the subject whatsoever, this method always bestows that which is wished for. Yoga meditation on a young woman causes people to become siddhas, there is no doubt of it. Just as grain is the secret essence of an ear of corn, or as the Sun's brightness manifests by its rays, or as the Moon's beauty is shown in falling rain, or as the earth becomes full of nectar by being watered, or as by seeing a flower one becomes filled with devotion, or through Mahadurga's prasad makes one a Lord of Siddhas, or as by the grace of Kula flower pleasure arises, or as remembering the Ganges frees from sin, so by this method of attraction one becomes like Shiva, and so meditating on a young woman gives boons. Therefore, always initiate the Nija Kauliki.

Bhairava is the rishi, ushnik is the metre, the devata is Devi Dakshina Kalika, it gives the fruit of the four aims of mankind. Purva is the bija, para is the Shakti.

Do limb (nyasa) and so forth using six long bijas, and the fourteen matrika vowels, each separately. Place them in the heart, on the hands and on the feet. Do the diffusion (nyasa) using the fifteen syllables of the root mantra. Meditate five ways as previously described.

Inside the lotuses and in the fifteen angles do pitha puja. There indraw and worship Devi Dakshina, adorned with Kula. Afterwards, worship Mahakala, then the pitha Shaktis Kali, Kapalini, Kulla in the first triangle; Kurukulla, Virodhini and Vipracitta; Ugramukhi, Ugraprabha and Pradipta; Nila, Ghana and Balaka; Matra, Mita and Mudrika. Outside this, from the east petal in order, worship Brahmani and the rest.

After doing thus, a pure person should recite the mantra and sacrifice daily. Reciting a lakh (times) at night is the essence of great purification.

One need have no other thought in this puja than that of a young woman. Reciting at night gives siddhi and one becomes Dakshina. Do limb nyasa, and after meditating on Devi, the wise man should recite the mantra. After worshipping the body of the cosmos using this method, one may attract heavenly, underworld and Naga maidens.

Worship Maha Kali in a forest, performing puja, meditation, application and recitation of the mantra. The Devi dwells equally in all these places. Daily recitation and the like has already been spoken of. When a person does forest puja of great maidens, it bestows purity.

Place a conch in the north east and draw a yantra there. Offer and practice on the eighth or fourteenth night (of the dark Moon). Initiated mortals should recite the mantra 108 times, naked, with their mouths full of tambula, with dishevelled hair, controlled of senses, eyes rolling with intoxication, in sexual intercourse with the supreme woman.

Worshipping at night, naked, using scent and flowers. adorned with Kula gems, she who is named in the strewn yantra, being the vira's beloved, is she who ought to be worshipped. After giving her wine, meat and the other substances of Kula sadhana, one should meditate on and offer to the guru.

Asking leave to dismiss her, (placing a flower) on the head, one should do the remaining actions. Dearest son! Do not do Kula puja without wine and flesh, else it destroys the good actions of 1,000 births.

Brahmins, in acts of subjugation, may offer honey in copper vessels instead of wine. Others should worship using Kula wine. This wine is drunk by yogis, this wine is the most excellent thing for yogis. For those for whom wine drinking is unsuitable, honey and sweet cakes may be offered.

Sixth Patala

Devi said: Dear son, now I speak to you of the rules of nyasa. One should invoke Devi as being diffused in the body, then doing nyasa. The best of sadhakas should first place

Devikuta at the top of the head. Draw a yantra using the previously declared rule, following the Kula path. Bow to the various pithas and using scent and blossoms worship Devi as Mahabhaga, the root Devi, with her attendants.

After reciting the mantra 100,000 times, then establish Oddiyana, then worship the pitha known as Yoganidra, where one should do puja to the indwelling ishtadevata 100,000 times. After going to Kamarupa, one should then worship Katyayani.

At night, reciting a mantra 100,000 times, one should then perform worship to Kamakhya. Going then to Jalandhara, firstly worship Purneshi. Reciting the mantra 100,000 times there too, then go to Purnagiri and again do puja and recite the mantra to Chandi. Entering Kamarupa, firstly worship Kamakhya. At the close, worship the Mahadevi Dikkaravasini.

At night, after reciting mantra to each of the Pitheshvaris in the seven pithas, then worship the ishtadevata. On completing the requisite number, then bow, saying "Devi, highest of the Kulas, in this act I, named such and such, of the gotra such and such and family such and such ask you, the desired goddess, to bestow the highest boon." If unsuccessful, one should do the previous actions again.

Otherwise, one may worship Mahishamardini in all the pithas. Then she becomes pleased, and bestows the Kula boon. On reciting the root mantra, one becomes a lord of all siddhi. One should go to a Rajavriksha tree, and worship the ishtadevata at its root, starting worship on a great night. The recitation should continue over the next three days. Deva, the best of sadhakas gains the fruit of 100,000 pithas by doing this.

Bhairava Maheshvara, if a sadhaka should recite the secret Mahishamardini root vidya, he may gain at will the siddhis Vetala, Khadga, Anjana and Tilaka.

Bhairava said: O Deveshi Chandika, if you love me, tell me how to obtain the great siddhi Vetala and the rest.

Devi said: The best of sadhakas, using Nimba wood should, on a Tuesday, at midnight, sit in sexual intercourse on a corpse. After digging a pit, he should recite the Mahishamardini (vidya) 800,000 times, then offering 1,000 times in the cremation ground.

Taking the ash, smear it on a staff and padukas, going to the cremation ground on a Durga eighth and offering libation there.

Doing puja according to rule on the corpse, a vira should then sit on the corpse and recite the mantra 1,008 times. O Natha, after giving animal sacrifice to the mother, he should then recite a mantra over the wood: "Sphrem Sphrem Mahabhaga Yogini, be lovingly pleased! Protect me with this staff I hold."

Whenever a Kaulika displays the staff and recites the mantra, he can pulverise whatever he wants and can conquer over and over again.

"Mahabhaga. Boon-giver, may these padukas go, go! May they travel 100 yojanas whenever I wear them!"

Taking copper and making a sword of 50 angulas length, and drawing a yantra on it, a person should recite the mantra. Sitting on a great corpse during a Kali day and reciting the mantra 1,000 times, (a sadhaka) should dig a pit in front of a Bija tree and do the binding and protection, reciting on a Kula eighth in the cremation ground at midnight. Firstly pleasing (the Devi), he should then do sacrifice in the cremation ground, using the three madhus with bilva leaf.

After the sacrifice, he should offer animal sacrifice to supreme Maya Devi Mahishamardini. Whosoever gives complete animal sacrifice to the great Unmukhi gains her favour. Dear son, say: "Take, wielder of the sword! Terrifying and fanged Maha Kali, true one of formidable form, Kam Im Um make! Kalyani, cut through my enemy!"

If a man should raise and strike with the sword, at the same time reciting the mantra, having cut, having cut and again having cut, he may achieve the act of sadhana.

Otherwise, he should sever the head of a male black cat with one blow, on a Tuesday, at the crossroads, at night, and should bury it with a mantra. Eating sacrificial food and emitting into a vessel, he should recite the mantra every night. Concentratedly reciting 1,008 times in the darkness, he should gaze at the dug-up vessel. Eating sacrificial food during the day, on a river bank, and immersing himself, he should wash it, reciting a mantra.

Dear son, I have spoken to you of opposition, through which a man can conquer. One should install and worship screaming Kalika, reciting the inimical Kali mantra 1,000 times. A mantrin then attains success in anjana-there is no doubt about this. A sadhaka, on making a powder of crushed bone, sandal, aguru and copper, worshipping according to rule, conquers all.

Kuleshvara, a man, worshipping Devi most attentitively at a Kula place, giving her Kula fish, Kula food and Kula wine. Reciting 1,008 times, a man may conquer all on earth using the Phut mantra. Even if the target is 100 yojanas distant, he may gain it. It comes to him from wherever it is on earth.

A man can contract his body, entering instantly a cleft, a small window, or a cavity. Dear son, Lord of Kulanathas, unless a man has the Durga or Kali mantra, the siddhis are concealed, there is no doubt of it.

Kularnava Tantra

The *Kularnava Tantra* is held in high esteem by Kaula tantriks. The phrase *Kularnava* means ocean of Kula. It focuses on the Urdvhamnaya, or Upper Tradition, the four others belonging to the four other faces of Shiva and representing the five elements of the tradition. This Tantra does exist in English translation (Prachya Prakashan, 1983) but this chapter is my own translation, done some years before. Each chapter is called an *ullasa* or bliss, this referring to the nectar of the Kaulas.

The essence of this third chapter is a eulogy about the magnificence of the Urdvhamnaya and its mantra, which is based upon the 21,600 breaths a human being takes each day and night. This upper form of Shiva is known as Ardhanarishvara, a form uniting both Shiva and Shakti in one form, representing the union of the outgoing and ingoing respirations (Ha and Sa).

A description (*dhyana,* meditation) of Ardhanarishvara, given in the same Tantra, is of some interest. "In the centre of the wood of Aeon Trees is a canopy formed from nine rubies.

Under the canopy is a throne studded with nine jewels. On the throne, on a triangular seat within the petals of the lotus, is Shiva, adorned with the Moon and the Sun with Devi Ambika forming half of his body..."

The mantras in *Kularnava Tantra* are all fashioned from the letters Ha Sa (shown in the picture above left as Hamsah). The Shri Prasada Para mantra is Hsaum, the Shri Para Prasada mantra Shaum, the Shri Prasada mantra Ham and the Shri Para mantra Sam.

Chapter Three

Shri Devi said: Lord of Kula, I want to hear about the most supreme of all dharmas, the Urdhvamnaya, its mantra and its magnificence. Speak to me!

Shri Ishvara said: Listen, Devi, I will declare that which you ask, by hearing which doctrine devatas become truly happy. I have never previously declared this, O Kuleshvari. Listen, dearest, I will tell you about the Urdvhamnaya out of my love for you.

Vedas, shastras, puranas are revealed but Shaiva and Shakta agamas are all renowned as being secret. The supreme secret of all secrets are the Kula Shastras, Parvati. Ambike, the Urdhvhamnaya is the supreme secret of all secrets, the very essence of the whole of the absolute (Brahman). Well hidden before, it is now revealed by me to you.

The five amnayas are born from my five faces are the east, the west, the south and the north and the upper, the five amnayas renowned for bringing liberation (Moksha). Many are the amnayas but there is none equal to the Urdvhamnaya. This is the truth, Vararohe, there should be no discussion about it.

Many distinctions have come from the secret four amnayas. In this Tantra, O Lady of the Kulas, I will first speak of them. O Kamini, many are those who study the four amnayas, but those knowing the Urvhamnaya knowledge are few.

Just as the myriad of particles of dust on the earth, so many are the mantras giving enjoyment and liberation from any one or other of the [four] amnayas. So too is there a myriad of lesser mantras from the amnayas, spoken by me to you for the grace of the world.

Sweetly smiling one, the devatas of all of these mantras originate from us both, giving the siddhis which are desired. I know every single one of these mantras and no one else. By my grace, only a handful of men out of millions are aware of them. If one knows just one amnaya, it bring liberation. How much more so for a person who knows all four. Clearly, he becomes Shiva.

The Urdhvamnaya is supreme, above the knowledge contained in the four amnayas. Whosoever desires it becomes a Siddha because of it. From the Urdvhamnaya comes all dharma. The Urdhvhamnaya is praised by all and there is nothing greater than it, it is said.

Kuleshani, a person is freed from the ocean of Samsara by the Urdvhamnaya. It is said that the heavens all worship the Urdvhamnaya. By practising it, Deveshi, one may become liberated in this very birth. The fruit of the Urdvhamnaya is greater than all other amnayas, the very quintessence. As in all worlds one should worship me, dearest, so in all amnayas one should worship the Urdvhamnaya.

Like Vishnu amongst gods, like the sun amongst heavenly lights, like Kashi amongst sacred bathing spots, like the philosopher's stone to gold and so forth, as Meru is to mountains, as the sandalwood is to trees, as Ashvamedha is to sacrifices, like a diamond is to pebbles, as sweetness is to other flavours, as gold amongst metals, as the cow is to quadrupeds, as the swan is to birds, as sadhu-ship is to the different ashramas, as a Brahmin is to the [other] varnas, as a king is to men, as the head is to the limbs, as musk is to the scents, as Kanchi is to cities, so of all paths the Urdvhamnaya is the best, Dearest.

In various births, the Urdvhamnaya is the fruit giver of supreme merit and not anything else, O Lady praised by Viras! Whosoever knows this Urdvhamnaya is wealthy, has Kula knowledge and is the best amongst men.

Not by vedas, agamas, shastras, puranas, expositions, sacrifices, penances, bathing in 10,000,000 sacred water, not by vows or by other things Deveshi, not by herbs or mantras previously declared can the Urdvhamnaya be known, but only from the mouth of Shrimad Guru.

Deveshi, one should immerse oneself in that all knowing ocean of compassion, having all characteristics, the knower of the essence of the Urdvhamnaya, the source of the Urdvhamnaya. Whosoever knows the Urdvhamnaya knows everything. All that is desired is achieved and the person is accomplished. This is true, true, O Varanana! Whosoever knows all this from the mouth of Shri Guru becomes liberated by the path of Kulashastra, there is no doubt of it.

Whosoever knows this amnaya, O Devi, has everything. He is worshipped, he is a true guru, he is a teacher, he is an astrologer, he is a mantrin, he is served by all, he is praised by all, he is a seer, he is Sattva Guna itself, he carries things through, he is a true doer of tapas, he is an achiever, he is a magician, he is a knower of veda, agama, shastra, all knowledge and so forth, he is the path, he is wise, he is a saint, he is a Kaulika, he is a true sacrificer, he is self-contained, he is a true reciter of mantra, he is a doer of magic, he is a yogi, he is a real accomplisher of acts, he is a hero, he is the quintessence, he is a good person, he is a knower of all, he is liberated, he is Shiva, dearest.

His family is pure, his mother is Lakshmi, his father is an accomplisher and may himself become liberated. His family line is praiseworthy in everything, his relatives and friends are complete. What is the use of many words? The Urdvhamnaya is the Supreme!

He who recites it gets the fruit of being greater than a king, remembered by all, famed, praised, visionary. Wherever he dwells, Shri Vijaya (Victory) dwells also. He is a true mendicant, without description, the essence of nectar.

The best type of person attains this Urdvhamnaya by the grace of Guru. Whosoever knows it, knows everything, is beloved, and becomes like me. The eastern amnaya is creation, the southern amnaya maintenance, the western amnaya dissolution and from the northern amnaya comes grace. The eastern amnaya is mantra yoga, the southern amnaya is bhakti yoga, the western amanaya karma yoga and the northern amnaya jnana yoga. There are 24 parts to the eastern amnaya, the southern has 25, it is said, the western amnaya has 32 parts and the auspicious northern amnaya has 36 divisions.

These divisions do not exist in the Urdvhamnaya, O lady of Kulas! Plainly, it is the very self of Shiva and no acts are known within it. I cannot speak of the magnificence of the Urdvhamnaya, from my love for you, you gave birth to this truth, Varanane!

So the greatness of the Urdvhamnaya is declared to you by me. Now, Kuleshani, the magnificence of the mantra is revealed. Its characteristics have not previously been spoken by me. I speak of them now because of your love. Listen, O One Dear to Me as Life Itself! The mantra called Shri Paraprasada is the chief mantra situated in the Urdvhamnaya. Whosoever knows this mantra, which is the supreme creator, becomes like Shiva.

Ha-Sa is the pathway breath takes in living creatures. This mantra exists in the form of exhalation and inhalation, dearest one. Just as clouds cannot exist without wind, and just as the sky is without limit, so the world cannot exist except by Shri Paraprasada mantra. The world of immovable and moving things comes from the Shri Paraprasada mantra.

Just as the absolute is without parts, Devi, and as wind moves the stem of the palm, and as the shoot is in the seed, and as sesame oil is in the sesame plant, and as brightness is in the sun and as the moon is in moonlight, and as fire is latent in wood, and as perfume is in flower, and as substances are dissolved in water, and as meaning is in sound, and as Shiva is in Shakti, and as ghee is in milk, and as colour is in fruit, and as sweetness is in sugar, and as coolness is in camphor, and as mantra favour exists in mantra punishment, and as divinity is in an image, and as movement is in the breeze, so also this universe exists in the Shri Paraprasada mantra. As in the fig tree, the subtle form of the tree subsists, so does the macrocosm exist in the Shri Paraprasada mantra.

Just as flavour exists in well-cooked food, Kuleshvari, and just as it is impossible to taste sweetness except by knowing saltiness, so those who are devoid of Mantra Shakti do not gain fruit or union with the Shri Paraprasada mantra. One should conceal the Shri Paraprasada mantra most carefully.

People wander in delusion amidst discussions, puranas, philosophies, amnayas, distinctions, mantras and shastras of various kinds, the 1,000 devas of the various texts, deluded by your Maya. They live and they die and have their share of the afflictions of Samsara, not knowing the Shri Paraprasada mantra, O Kuleshvari.

They do not gain liberation, deluded by your Maya. Whosoever knows and is devoted to the lovely form of Shri Guru gains the Shri Paraprasada mantra and having known this becomes completely liberated. Whosoever in 1,000 previous births knew Shaiva and the other (paths) and worshipped the mantras of the four amnayas will come to sing of the Guru. Whosoever knows the Shri Paraprasada mantra comes to be liberated from evil limitations, a purified being, dear to the Guru, and not in any other way.

Brahma, Vishnu, Rudra, Indra and so forth; various heavenly beings, the Vasus, the Rudras, the Sun, the dikpalas, Manu, the Moon and so forth; Markandeya and the other munis, Vashishta and other lords of saints, Sanaka and other lords of yoga, liberated beings like Shuka and so forth; yakshas, kinnaras, gandharvas, siddhas, fairy folk and others praise the one who attains this meritorious Shri Paraprasada mantra, from which comes all results.

One like this has everything, is worshipped, gains knowledge, is effulgent, happy, free from disease and from recitation of the Shri Paraprasada mantra gains kingship, heaven and liberation. Also the knower of this mantra exhausts all karma and attains to the realm of Brahma, Indra, Rudra and Vishnu. Whoever attains to this becomes happy in all, and a true person.

This is the wish-fulfilling gem, the heavenly cow (Kamadhenu), in the palace in the centre of the Aeon Trees.

By reciting the Shri Paraprasada mantra, even Lord Kubera becomes a servant. Just as by using the philosopher's stone, iron becomes gold, so by the recitation of the Shri Paraprasada mantra, a beast (pashu) becomes a lord of beasts (Pashupati). Whosoever knows the Shri Paraprasada mantra has everything. He knows the best mantra of all and becomes like You and Me.

O Parvati, one who knows this mantra core of Shri Paraprasada mantra, even if a cooker of dogs, becomes liberated. If even a dog-eater becomes liberated by knowing the mantra essence of Shri Paraprasada mantra, how much more so one who knows its practice? The knower of the Shri Paraprasada mantra may do whatsoever he wishes, if, having heard that mantra he should meditate on it. This person, O Maheshani, previously initiated according to the genuine line (parampara) and knowing the Shri Paraprasada mantra, becomes one with So-Aham, there is no doubt. The knower of Shri Paraprasada mantra in the body is established eternally in the 14 worlds, aware of both moveable and immovable equally.

Wherever he who knows the Shri Paraprasada mantra resides, that place is like a celestial palace of broad dimensions. O Lady of Kulas, the knower of the essence of Shri Paraprasada mantra is praised by Gods and Asuras, over and over again. The place where the knower of Shri Paraprasada mantra stands is a place of the Siddhas, thronged with hosts of saints and devas. A knower of the Shri Paraprasada mantra knows all mantras, whether they be Shaiva, Vaishnava, Daurga (of Durga), Arka (of the Sun), Ganapatya (of Ganesha) or Lunar. The sight of one who has the Shri Paraprasada mantra on the tip of his tongue, liberates even a cooker of dogs. Whether a Brahmin or any other case, whether pure or impure, he who recites Shri Paraprasada mantra is liberated, there is no doubt of this.

Whether moving or staying still, whether awake or asleep, this Shri Paraprasada mantra always gives results, O Deveshi. The other thousands of mantras each give one single result but, Kuleshi, this king of mantras swiftly gives every kind of result. Of all mantras, this Shri Paraprasada mantra is the highest. Whether ignorant or knowledgeable, this mantra always gives fruit. (Like the pairs) Sachi and Indra, Rohini and the Moon,

Svaha and fire, light and the Sun, Lakshmi and Narayana, Saraswati and Brahma, night and day, fire and the moon, bindu and nada, Prakriti and Purusha, that which is and that which is not, enjoyment and liberation, Prana and Apana, word and meaning, that with parts and that without parts, happiness and sorrow and so forth and in all complements seen and heard by me, this (Shri Paraprasada mantra) exists in all, there is no doubt. This Shri Paraprasada mantra exists in all male and female forms and is therefore the self of all, O Kuleshvari.

Kuleshvari, it is attained by meditation, is without form, the supreme absolute, without divisions, pure, eternal, without qualities, all pervading, the eternal imperishable thing, whose sphere of action is outside the realms of speech and mind. One becomes illuminated by virtue of the sadhana of Shri Paraprasada mantra. Devi, this mantra is, therefore, known as Shri Paraprasada mantra, the supreme quintessence, having the nature of truth-consciousness-bliss, Shiva and Shakti, bestower of both enjoyment and liberation, with and without karma, with and without qualities (gunas).

This Shri Paraprasada mantra is the chief of all mantras. By reciting it, one gains both enjoyment and liberation, of this there is no doubt. What use here of much speech? It is the all-essence!

Listen, dearest. There is no mantra equal to the Shri Paraprasada mantra. It is supreme knowledge, it is supreme tapas, it is supreme meditation, it is supreme worship, it is supreme initiation, it is supreme mantra recitation, it is the highest of all tattvas, it is the greatest of all vows, it is the best of sacrifices, it is the highest of the high, it is supreme bliss, it gives the best result, it is the very absolute (parabrahman), it is the highest refuge, it is the most secret of all secrets. After knowing this quintessential and boon giving mantra, one always becomes steady, Dear One.

Employing the rules of sadhana using the injunctions of the agama, one should recite the Shri Paraprasada mantra 108 times and one becomes liberated from the five great sins such as killing Brahmins and so forth.

Bibliography

Agrawala, S. V.: *The Glorification of the Great Goddess*, All-India Kashray Trust, Varanasi, 1963.

Anand, Margo: *The Art of Sexual Magic*, G.P. Putnam's Sons, New York, 1995.

Arthur, A.: *Tantrik Texts: Kalivilasa Tantra*, Luzac and Co., London, 1917.

Arvind and Shanta: *Tantra: The Secret Power of Sex*, Jaico Publishing, Bombay, 1976.

Arvind, S.: *Tantra: The Secret Power of Sex*, Jaico Publishing, Bombay, 1976.

Bagchi, P. C.: *Studies in Tantras*, University of Calcutta Press, Calcutta, 1939.

Baird, Robert: *Religion in Modern India*, Manohar, New Delhi, 1981.

Banerjea, J. N.: *Pauranic and Tantric Religion*, University of Calcutta, Calcutta, 1966.

Basu, Manoranjan: *Tantras: A General Study*, Shrimati Mira Basu, Calcutta, 1976.

Benoytosh, A.: *Guhya Samaja Tantra*, Gaekwad's Oriental Series, Baroda, 1931.

Bhattacharjee, A.M.: *Hindu Law and the Constitution*, Eastern Law House, Calcutta 1994.

Bose, D. N.: *Tantras: Their Philosophy and Occult Secrets*, Kali Press, Calcutta, 1965.

Chattopadhyaya, Sudhakar: *Reflections on the Tantras*, Motilal Banarsidass, New Delhi, 1978.

Chawdhri, L. R.: *Pràcticals of Mantras Tantras*, Sagar Publications, New Delhi, 1985.

Christopher, S.: *Tantra: The Yoga of Sex*, American Oriental Series, New Haven, 1974.

Colin A. Cole, *Asparsa-yoga: a Study of Gaudapada's Mandukya Karika*, Motilal Banarsidass, Delhi, 1982.

Cozort, Daniel: *Highest Yoga Tantra*, Snow Lion Publications, New York, 1986.

Daniel, C.: *Esoteric Teachings of the Tibetan Tantra*, Snow Lion Publications, New York, 1986.

Datta, V.N.: *Sati: Widow Burning in India*, Manohar, New Delhi, 1990.

Dehija, Vidya: *Yogini Cult and Temples: A Tantric Tradition*, National Museum, New Delhi, 1985.

Donat, K.: *Meditate the Tantric Yoga Way*, George Allen and Unwin, New York, 1978.

Doniger, Wendy: *Textual Sources for the Study of Hinduism*, Barnes and Noble, New Jersey, 1988.

Dutt, Romesh C.: *The Ramayana and the Mhabharata,* J. M. Dent & Sons, London, 1910.

Elizabeth, A.: *Kali: The Black Goddess of Dakshineshwar*, Nicolas-Hays, New York, 1998.

Frawley, David: *Tantric Yoga and the Wisdom Goddesses*, Passage Press, Salt Lake City, 1994.

Garrison, Omar: *Tantra: The Yoga of Sex*, Avon Books, New York, 1964.

Gavin and Yvonne: *Tantric Yoga: The Royal Path to Raising Kundalini Power*, Samuel Weiser, New York, 1989.

George, Christopher. S.: *The Candamaharosana Tantra*, American Oriental Series, New Haven, 1974.

Ghurye, G.S.: *Indian Sadhus*, Popular Prakashan, Bombay 1995.

Gopinath, K.: *Aspects of Indian Thought*, University of Burdwan, Barakar, 1986.

Govinda, Lama Anagarika: *Foundations of Tibetan Mysticism*, Rider and Company, London, 1960.

Gupta, Sanjukta: *Hindu Tantrism*, Brill, Leiden, 1979.

Gurumurthy, S.: *Hindu Heritage, Assimilative, Not Divisive.* Vigil, Madras 1993.

Haich, Elizabeth: *Sexual Energy and Yoga*, Aurora Press, New York, 1982.

Harman, William: *The Sacred Marriage of a Hindu Goddess*, Indiana University Press, Bloomington, 1989.

Hopkins, Jeffrey: *Kalachakra Tantra Rite of Initiation*, Wisdom Publications, Boston, 1982.

Isaeva, V.: *Shankara and Indian Tantric*, SUNY, New York, 1993.

Iyengar, B.K.S.: *Light on the Yoga Sutras of Patanjali*, Aquarian Press, London 1993.

John R. and Eric J. Sharpe: *Hinduism,* Oriel Press Limited, England, 1972.

John. Stratton: *Krishna, the Butter Thief*, Princeton University Press, Princeton, 1983.

Keay, John: *India: A History*, MacMillan, New York, 2000.

Kenneth W: *The Religion of the Hindus,* The Ronald Press Company, New York, 1953.

Khanna, Madhu: *The Tantric Way: Art, Science, Ritual*, Thames and Hudson, London, 1977.

Klostermaier, Klaus K.: *A Survey of Hinduism,* State University of New York Press, Albany, 1989.

Maharishi, Yogiraj Vethathiri: *Journey of Consciousness*, MacMillan Limited India, Delhi, 1992.

Mahendranath, S.: *Comparative Studies in Vedaantism*, Humphrey Milford, Bombay, 1927.

Natarajan, B.: *Thirumandiram: A Classic of Yoga and Tantra,* Babaji's Kriya Yoga Publications, Montreal, 1993.

Olson, Eleanor: *Tantric Buddhist Art*, China House Gallery, New York, 1974.

Omar, G.: *Tantric Yoga: The Royal Path to Raising Kundalini Power*, Avon Books, New York, 1964.

Pandit Hemchandra: *Kamaratna Tantra*, Assam Government Press, Shillong, 1928.

Pandit Shiv: *Yoga and Sex*, B. I. Publications, Bombay, 1973.

Pushpendra, K.: *Hindu Religion and Ethics,* Asian Publication Services, New Delhi, 1997.

Rajneesh, Bhagwan: *Tantra: The Supreme Understanding*, Chidvilas, Boulder, 1993.

Rochford, E. B.: *Hare Krishna in America,* Rutgers University Press, New Brunswick, 1985.

Saraswati, Swami Janakananda: *Yoga, Tantra and Meditation*, Random House, New York, 1974.

Swami Janakananda: *Yoga, Tantra and Meditation in Everyday Life*, Rider Books, London, 1975.

Teun, G.: *The Candamaharosana Tantra*, Motilal Banarsidass, Delhi, 1985.

Troy W.: *Hinduism: Its Historical Development,* Barron's Educational Series, New York, 1974.

Vivekananda, Swami: *Raja Yoga,* Ramakrishna-Vivekananda Center, New York, 1973.

Wentz, W. Y.: *Tantric Yoga and the Wisdom Goddesses*, Oxford University Press, London, 1958.

White, Meryl: *Karmapa: The Black Hat Lama of Tibet*, Luzac and Company, London, 1976.

Yogananda, Paramahansa, *Tantra: The Supreme Understanding,* Self-Realization Fellowship, Los Angeles, 1946.

Zimmer, Heinrich: *Yogic and Tantric Medicine*, Joseph Campbell, Delhi, 1946.

Index

T

U

V

W

Y

❑❑❑